Metaphors
for Environmental
Sustainability

Metaphors for Environmental Sustainability

Redefining Our Relationship with Nature

BRENDON LARSON

Yale

UNIVERSITY PRESS

New Haven and London

Published with assistance from the Mary Cady Tew Memorial Fund.

Yale University Press books may be purchased in quantity for educational,
business, or promotional use. For information, please e-mail sales.press@
yale.edu (U.S. office) or sales@yaleup.co.uk (U.K. office).

Set in Adobe Minion type by Duke & Company, Devon, Pennsylvania.
Printed in the United States of America by Sheridan Books, Ann Arbor,
Michigan.

Library of Congress Cataloging-in-Publication Data
Larson, Brendon.
Metaphors for environmental sustainability : redefining our relationship
with nature / Brendon Larson.
p. cm.
Includes bibliographical references and index.
ISBN 978-0-300-15153-4 (hardcover : alk. paper) 1. Nature—Effect of
human beings on. 2. Sustainability. 3. Metaphor. I. Title.
GF75.L37 2011
304.2—dc22
2010035174

A catalogue record for this book is available from the British Library.

This paper meets the requirements of ANSI/NISO Z39.48–1992
(Permanence of Paper).

10 9 8 7 6 5 4 3 2 1

To the natural world and human experience of it
in seven generations

Criticism of metaphoric worlds, or visions,
becomes one clear and important—perhaps the clearest
and most important—instance of a general
human project of improving life by criticizing it.
—Wayne Booth, "Metaphor as Rhetoric"

Technical language . . . is, in the last resort,
an efflorescence of every-day language.
—Ludwig von Bertalanffy, "An Essay on the
Relativity of Categories"

The ecological crisis is one primarily and fundamentally of
metaphor. . . . The resolution of the ecological crisis depends,
then, upon the extent to which life-giving metaphors
can be restored to our communal life.
—Richard Underwood, "Toward a Poetics of Ecology"

Contents

Preface

We don't see things as they are. We see them as we are.

—*Anaïs Nin,* The Diary of Anaïs Nin

The way we speak about the natural world is not a transparent window, because it reflects the culture in which we live and its priorities and values. In the discourse about sustainability, for example, we look to environmental science for the facts, often neglecting the value-laden language in which they are communicated. I was initially trained in evolutionary ecology, then changed fields to examine the confluence of facts and values in our thinking about the environment. I had become immersed in scientific concepts but, to a large extent, had forgotten—as it is so easy to do—that they represent a complex way of human knowing. They are not nature itself.

Metaphor was the entrée to this realization, and a formative moment occurred when I was discussing a paper on invasive species with a group of students. Though I had extensive experience with these species, I had always been inside the worldview of conservation biology. Its metaphors had been imperceptible, and their presence and value-ladenness suddenly struck me. It seemed that scientists were not being objective in the way many people assume they are supposed to be; instead, they were effectively making political claims. I thus began to examine how the metaphors of environmental science interweave science and society and the implications of this interweaving, themes that have fascinated me ever since and have given rise to this book. My examination of these themes has convinced me of the necessity of considering both material issues, such as biodiversity loss, as well as contributive worldviews.

With this book, I hope to convince you of the importance of thinking about the social context of the metaphors of environmental science, if not more widely. Intellectually, this exploration will lead us into critiques of some of the distinctions—such as between facts and values and science and society—that underlie much of how we often think about science. In this respect, I engage with the possibility of what has been called postnormal science, a form of science that more explicitly engages with society and its values in the interests of sustainability. This work therefore represents a foray into the linguistic dimensions of such a science. It is quite normative and even subversive. I not only seek to provide a better understanding of how our environmental metaphors operate in context, but also propose that in some instances we need to reframe them so that they are more consistent with values rooted in sustainability. I considered excluding these proposals, but they are closer to the heart of the matter for me.

Although many scholars have written about the function of metaphor in scientific inquiry, relatively few have written about their socioethical implications, and their writings occur as scattered journal articles and book chapters on specific case studies. The closest book-length treatments to this one are technical or focused on the medical realm, whereas this one aims to welcome a number of audiences interested in science, metaphor, and/or sustainability. I have written this book for those who might benefit from this rumination on metaphoric choices, including reflective environmental scientists (focusing on conservation biologists, ecologists, and evolutionary biologists, though the ideas are applicable more broadly), everyday citizens with an interest in science and its potential contribution to sustainability, and perhaps even policymakers. It will also appeal to readers with a general interest in metaphor and/or environmental sustainability, two popular themes for contemporary readers. Although my focus has not been a particular technical audience, I hope that this inquiry will also make some small contribution within fields such as ecocriticism, ecolinguistics (and other fields devoted to critically examining metaphors), environmental ethics, public understanding of science, science communication, and linguistic/ rhetorical science studies. It would be appropriate for upper-level courses in any of these fields, particularly in courses that examine interdisciplinary links between environmental science and society (e.g., environmental studies). The risk of an interdisciplinary and synthetic project such as this one is that some linguistic specialists and students will critique my interpretation of metaphor just as some sociologists will discount my method, philosophers my discussion of metaphoric facts and values, and biologists my focus on the social dimensions of their science. I depend on them to pick up where I have left off.

Acknowledgments

I owe thanks to innumerable people for their contributions along the way, but especially to Jim Proctor for being a model for me intellectually, especially in his dedication to interdisciplinarity. Leaving biology for an uncertain interdisciplinary future was an agonizing process, and I appreciate four biologists, Alice Alldredge, Bruce Milne, Joshua Schimel, and Bruce Tiffney, for encouraging my interest in the socioethical dimensions of biological knowledge. I wish to thank the following colleagues who provided stimulating discussion and/or generous encouragement at various points along the way: Glenn Adelson, Charles Bazerman, Paul Chilton, John Du Bois, Scott Gilbert, James Griesemer, Paul Griffiths, George Lakoff, John Lyne, Michael Osborne, Steward Pickett, Massimo Pigliucci, Hilary Putnam, O. J. Reichman, Wade Clark Roof, Ricardo Rozzi, Karola Stotz, Alan Wallace, and Mark Westoby. To the other scholars I cite in these pages—as well as the authors, poets, and thinkers whom I don't, yet who influenced me—thank you for providing material for inspiration and reflection born of your own intellectual odysseys.

I would like to acknowledge those who took part in a working group on ecological metaphors that I organized with funding from the National Center for Ecological Analysis and Synthesis (NCEAS), a conversation that helped to kick-start an early stage in the development of these ideas: Ken Baake, James Bono, Kim Cuddington, Yrjö Haila, Stuart Hurlbert, Evelyn Fox Keller, Denise Lach, Gregory Mikkelson, Tim Rohrer, Char Schell, and Larry Slobodkin.

I am grateful to the following colleagues and friends who took the time to provide constructive feedback on portions of an early draft of the manuscript: Ken Baake, Stephen Bocking, Chet Bowers, Theodore Brown, Roslyn Frank, Jarmo Jalava, Evelyn Fox Keller, Jozef Keulartz, Brigitte Nerlich, Bryan Norton, Peter Taylor, Andrew Thompson, Gerry and Rosemary Waldron, and several anonymous reviewers. I particularly appreciate insightful comments from Adam Dickinson, Kevin Elliott, Andrew Goatly, David Harrison, Colin Milburn, Betty Smocovitis, and Cor van der Weele that necessitated a rethinking of some sections. Although I have woven the web of concepts in these pages with the assistance of all these people, I remain responsible for errors and shortcomings that remain.

I am indebted to the supportive interdisciplinary and collegial atmosphere in the Department of Environment and Resource Studies within the Faculty of the Environment at the University of Waterloo, including a sabbatical during which I completed the manuscript. Joe Bevan helped to prepare the figures. I wish to thank Jean Thomson Black, my editor at Yale University Press, for believing in this book through the entire process of shepherding it into being, and Jaya Chatterjee and Ann Twombly for their careful editorial work.

Finally, I treasure the love and support of my parents, Dale and Mary Elizabeth; my sisters, Mary Rie, Sacha, Michaila, and Tamara; my radiant daughter, Kyra Kestrel; *et mon amoureuse,* Karolyne Natasha.

I have adapted parts of the case studies here from previously published materials. A few sections of chapters 2 and 3 are revised from "The Social Resonance of Competitive and Progressive Evolutionary Metaphors," *BioScience* 56:997–1004, 2006. Scattered portions of chapter 5 are revised and expanded from "DNA Barcoding: The Social Frontier," *Frontiers in Ecology and the Environment* 5:437–442, 2007. Parts of chapters 5 and 6 have been updated from "Should Scientists Advocate? The Case of Promotional Metaphors in Environmental Science," in *Communicating Biological Sciences: Ethical and Metaphorical Dimensions,* ed. B. Nerlich, R. Elliot, and B. M. H. Larson (Burlington, VT: Ashgate, 2009). Several paragraphs in chapter 6 have been adapted from "The War of the Roses: Demilitarizing Invasion Biology," *Frontiers in Ecology and the Environment* 3:495–500, 2005, and from "Who's Invading What? Systems Thinking about Invasive Species," *Canadian Journal of Plant Sciences* 87:993–999, 2007.

I owe thanks for Figure 9, courtesy of K. David Harrison; Figure 10, courtesy of Robert Dooh; Figure 12, from Dean MacAdam; and Figures 11 and 16, courtesy of James Kamstra. I also thank the following individuals and publishers who granted permission to reprint excerpts of their work herein: Pablo Neruda, *The Book of Questions.* Copyright © Fundación Pablo Neruda, 2010. English translation copyright © 1991, 2001 by William O'Daly. Reprinted with permission of Copper Canyon Press, www.coppercanyonpress.org. Pablo Neruda, *Estravagaria.* Copyright © Fundación Pablo Neruda, 2010. English translation copyright © 1974 by Alastair Reid. Reprinted by

I
Metaphor and Sustainability

*Even though it is the demise of earthly forests that elicits our
concern, we must bear in mind that as culture-dwellers
we do not live so much in forests of trees as in forests of words.*
—Neil Evernden, The Natural Alien

*We are using the wrong language. . . . We have a lot of genu-
inely concerned people calling upon us to "save" a world which
their language simultaneously reduces to an assemblage of
perfectly featureless and dispirited "ecosystems," "organisms,"
"environments," "mechanisms," and the like. It is impossible
to prefigure the salvation of the world in the same language
by which the world has been dismembered and defaced.*
—Wendell Berry, Life Is a Miracle

Figure 1. A springtime walk in the woods.

cool spring breeze caresses my skin as I wander through a carpet of spring wildflowers beneath towering sugar maple trees (Figure 1). After a long Canadian winter, I yearn for the spring—the growth and flowering of bloodroot and hepatica, and then trilliums and violets—a virtual frenzy of renewed life. At times, I focus on the particulars, on this queen bumblebee buzzing through the understory or a particular shoot of jack-in-the-pulpit as it emerges from the damp earth. At such moments, I feel an intimacy with the natural world. Though knowing the identities of species may impede this direct experience, it also breeds a familiarity that comes from years of observing the cycles of the seasons.

I have been a naturalist for as long as I can remember, yet I also have been trained as an ecologist. In this mode, I supplement observational study with a search for patterns in nature. Many environmental scientists have followed this path; an exemplar is Robert MacArthur, whose naturalistic observations of warblers foraging in treetops led to theories of competition. Thus, walking through these same woods, I consider how the species in this community colonized and assembled through competition, mutualism, catastrophic disturbance, and drift. I ask whether any of its species are ecosystem engineers, alien, or invasive. Taking an evolutionary standpoint, I wonder about genetic mechanisms such as gene flow and natural selection, and ask questions about evolutionary fitness.

In the midst of human-caused environmental changes, I might turn to questions that conservation biologists ask. What is the level of ecosystem health and stability of this forest? Has fragmentation affected it? What ecological functions and services does it provide? Is it a biological hot spot? Does it contain any keystone or umbrella species? Is it a source or a sink for local species? Such questions often address fundamental issues of whether we should conserve, manage, or restore it.

Metaphors inform nearly all these investigations and questions (see Table 1), which is why I have devoted myself to studying them and how they influence our perception of the natural world. Although other linguistic elements shape scientific facts and lend them persuasive ability and social authority, scholars have emphasized the importance of metaphor. In their histories of ecology, for example, historians often remark on the commonness of metaphors.[1] We use metaphors all the time, as a moment's consideration of this very sentence will reveal. It refers to our "use" of metaphors as if they are tools and to thoughts as something "revealed" as if by sight. The frequency of metaphors should not be surprising. We often rely on metaphors in our attempts to understand the world around us: they allow us to conceptualize something of one sort as if it were another. My examples also demonstrate the implicit comparison of a metaphor, whereas similes state more explicitly that something *is like* something else.

Although scientists have traditionally denounced metaphor as imprecise rhetorical embellishment, diverse scholars have attested to the essential role of metaphors in science— a claim illustrated in my opening vignette. For example, in his book *Making Truth: Metaphor in Science,* the chemist Ted Brown of the University of Illinois argued, "Metaphorical reasoning is at the very core of what scientists do when they design experiments, make discoveries, formulate theories and models, and describe their results to others." His view echoes the Harvard evolutionary biologist Stephen Jay Gould's assertion that metaphors are "indispensable in human reasoning."[2] Historians and philosophers also emphasize the role of metaphors in scientific thought and practice, demonstrating that they may be extended into theoretical analogies and models that allow prediction and empirical testing.[3] Metaphoric ambiguity may actually be a fertile source of inquiry, and metaphors may allow leaps beyond stabilized meanings that revolutionize the process of scientific exploration. Science is built on metaphorical foundations, so metaphors occur in much of the science we hear about each day. Thus, we require greater sensitivity to their presence and implications.

Metaphor is a key element in scientific inquiry because it enables us not only to understand one thing in terms of another but also to think of an abstraction in terms of something more

Table 1. A Partial Who's Who of Metaphors Used in Environmental Science

You will recognize many of the metaphors in this list, as they appear not only in the pages of primary scientific research articles (from which I have collected them), but in everyday language as well. They are sorted into nonmutually exclusive categories of conservation science. Not all of them would qualify as feedback metaphors as defined in the text.

Ecology
assembly rules
biodiversity
biotic resistance
clone
colonize
community
competition
competitive exclusion
cooperation
disturbance
ecological integrity
ecosystem engineer
efficiency
environmental noise
equilibrium
evenness
food chain
food web
guerrilla
guild
interaction strength
niche
parasite
patch
phalanx
population
productivity
propagule pressure
range
species richness
stability
succession
territory
trophic cascade
understory

General
function
management
mechanism
restoration
sustainability

Types of Species
alien
exotic
flagship
hybrid
invasive
keystone
native
natural enemy
naturalized
umbrella

Systems Theory
chaos
complexity
hierarchy
network
node
redundancy
resilience
scale
self-organization

Conservation Biology
biodiversity hot spot
biodiversity levels
biological inertia
biological legacy
catastrophe
connectivity
conservation
conservation portfolio
corridor
ecological trap
ecosystem health
ecosystem service
extinction debt
fragmentation
global warming
invasional meltdown
island biogeography
natural capital
protected area
source-sink dynamics

Conservation Genetics
adaptive radiation
bottleneck
DNA barcoding
fitness
founder effect
gene flow
genetic drift
mutation

Taxonomy
class
family
kingdom

concrete and everyday. According to cognitive linguists, this association between abstract and experiential realms often results from a conceptual mapping that develops during childhood. Our metaphors, and therefore our reasoning, derive from our bodies; they are embodied. As the linguist George Lakoff and philosopher Mark Johnson, authors of the classic exposition of this idea, *Metaphors We Live By,* explain, "Human concepts are not just reflections of an external reality, but . . . they are crucially shaped by our bodies and brains, especially by our sensorimotor system."[4] Metaphors are not only linguistic; they are thoroughly conceptual. We are neither bats nor butterflies; instead, we experience and interact with the world as human beings. We use this basic experience of the world to interpret novel phenomena and the unknown.

For example, we are all familiar with the sense of bodily balance. Cognitive psychologists have shown that this experience provides a basis for conceptualizing other domains in our lives, including balance within our personalities and working lives, the justice system, the international political arena—and balance and beneficent order in nature. Ecologists may reject the "balance of nature" as a vulgar popularization, yet, after careful examination, one scholar concludes that this metaphor constitutes ecological theory rather than being "an imprecise precursor of the theoretical concept of mathematical equilibrium."[5] As beings for whom balance matters, perhaps we cannot help questioning whether our environment is in balance. This metaphor also allows us to draw on shared experiences of bodily balance to communicate with one another.

Metaphors also help us interpret the novel and the unknown by invoking our shared cultural context.[6] Biologists thus speak of aliens, arms races, and fitness, as well as genetic blueprints, codes, and information. And we conceptualize the social lives of the bumblebees I observed this spring as worker-queen hierarchies. These examples demonstrate how metaphors wed our search for empirical knowledge to rhetoric. A metaphor must be both accurate and comprehensible if people are to understand and embrace it. And because scientists and non-scientists live in a similar bodily and cultural context, well-chosen metaphors appeal to both audiences. Whether from embodied or cultural sources, metaphors provide a remarkable way for scientists to describe their ideas and findings to others, including colleagues, scientists in other disciplines, and even nonscientists.

In this manner, metaphors create numerous links between science and society. Ecologists attribute a specific meaning to terms such as *community, competition, disturbance, equilibrium, invasive, native, stability, succession,* and *territory,* for example, but these terms still maintain elements of their ordinary meaning. These multiple meanings, which vary with context, are called *polysemy.* Scientific texts may narrow polysemy, but in popular contexts metaphors will be continuously open to alternative meanings. Thus, the use of such terms stymies attempts to keep technical and ordinary meanings distinct. This mixing of science and society, and its implications, is the focus of this book.[7]

Before continuing, I want to dispense with the idea that metaphors we have integrated fully into our discourse are insignificant because they are "dead." Briefly consider the life cycle of a metaphor. Novel metaphors, such as those used commonly in poetry, have not appeared before, so we must expend quite a bit of effort to comprehend them when we first encounter them. With continued use, however, these metaphors eventually become hackneyed.[8] Although we may no longer recognize their origin, these dead metaphors may still have conceptual significance, ranging from very small (the leg of a table) to very large (the metaphor of mechanism, discussed below). Their literalization by no means erases their origins, as a little linguistic archaeology reveals, and they may in fact be all the more influential because they are unconscious and no longer recognized as metaphorical. Increasingly, they might be mistaken for reality itself. They may still be metaphors by which we live.

Stated differently, even dead metaphors do something. In technical parlance, they are performative. Because they influence our conception of the world, they catalyze particular outcomes. They dramatically affect our worldview, which is not just our view of the world but also our way of living within it. Thus, the metaphors we select have very real outcomes in the constitution of culture and in the political realm. As the Harvard historian Anne Harrington succinctly put it, metaphoric choices create new connections between domains, leading to "crossroads, the traversing of which has implications for the future." A compelling example comes from accounts by Susan Sontag, the American author and literary theorist, and other scholars of how medical metaphors engender approaches to disease that tend to marginalize women's conceptions of their own bodies and their own health. This is why the American developmental biologist Scott Gilbert declared, "Humanity must be made safe from metaphor." Throughout this book, I seek metaphors for the natural world that will keep it safe, thereby developing a theme introduced by numerous previous scholars.[9]

Briefly consider a metaphor that plays a significant role in how we live our daily lives: Time Is Money. We often speak of time as if it were money—for example, in everyday expressions such as "You're wasting my time," "This device will save you hours of work," "How will you spend your weekend?" and "I've invested a lot of time in this relationship." Every metaphor brokers what is made visible or invisible; this one highlights how time is like money and obscures ways it is not. Time thus becomes something that we can waste or lose, and something that diminishes as we grow older. It is abstracted in a very linear, orderly fashion. This metaphor, however, fails to disclose important phenomenological aspects of time, such as how it may speed up or slow down, depending on our engagement with what we are doing. We may instead conceive of time as quite fluid—as a stream, for example—though we lose sight of this to the extent that we have adopted the worldview of Time Is Money. Summarizing its influence, Lakoff and Johnson concluded, "The Westernization of cultures throughout the world is partly a matter of introducing the Time is Money metaphor into those cultures."[10] By framing our relationship to an abstract entity in a specific way, such a metaphor contributes to a particular way of being and acting in the world.

Communicating Sustainability

This book specifically considers the role of metaphors used in environmental science in how we think about sustainability. I will discuss sustainability at greater length later because it is itself a critical metaphor in what follows, but for the time being, let us think of it as the question of whether our actions will impede or maintain the Earth's ability to support both our lives and those of other species in the future. We face diverse challenges in attaining this objective, from biodiversity loss to global warming, yet nearly all of them are couched in metaphors that straddle science and society. Even the basic questions we use to frame these challenges are enshrined in such metaphors: how are we to respond to the environmental "crisis" and its threat to our "life-support system"?

Consider two metaphors that environmentalists have adopted to encourage a more sustainable human presence on our planet: the Earth as a goddess or as a machine.[11] The former, Gaia, derives from the name of the Greek Earth goddess. It was introduced in a scientific context to describe the remarkable stability of Earth's atmosphere, its self-regulatory capacity as a unitary whole composed of both living and inorganic parts. By contrast, spaceship Earth provides an image of our planet as a finite system hurtling through endless space, encouraging us to better manage its resources. With a little reflection, it is easy to grasp the contrast between these metaphors. The former personifies the Earth; the latter mechanizes it. The former places the locus of control in the Earth itself; the latter places it with human technocrats. I'm sure you can come up with other contrasts, but the pivotal question is whether the metaphors we have chosen will help us on the path of sustainability or lead us further astray.

When we personify the Earth as Gaia, we might begin to see and appreciate *her* as a holistic living being on whom we are utterly dependent for our home. This dependence was a revolutionary insight for some people around the time that Gaia was proposed, yet its holistic connotations also contributed to many scientists' rejection of this imagery. Some feminists, too, pointed out that Gaia is too easily objectified as Mother Nature, which not only perpetuates problematic associations between nature and women—that is, that she is either a kind lover or a capricious and judgmental mother—but also contributes to the myth that we do not need to intervene because Gaia can self-regulate: Mother Nature knows best.[12]

Some scientists prefer the image of spaceship Earth, one that perhaps suggests we should leave environmental decisions to expert technocrats. These experts can provide mechanistic solutions for the environmental problems we face. My focus will be on this expert vision, because we rely so much on science in thinking about the environment. It might seem that this capitulates to its predominance in environmental politics, but instead it acknowledges its very real significance.

I wish to explore what might otherwise remain hidden within our scientific metaphors related to the environment, especially in field-based biological and conservation sciences. This is a thorny task: what is hidden is partly our assumption that we must rely on science for environmental solutions because it is purely empirical and neutral. But this ignores broader questions about the role of science in environmental affairs. In asking such questions in the context of metaphor, I have been influenced by the MIT scholar Donald Schön's admonition: "The essential difficulties in social policy have more to do with problem setting than with problem solving, more to do with ways in which we frame the purposes to be achieved than with the selection of optimal means for achieving them."[13] It is important for the metaphors that environmental scientists choose to contribute solutions to the correct problems, rather than to ones that have been ill-defined.

The issue here is that environmental scientists primarily evaluate the epistemic dimension of environmental metaphors, that is, their aptness for describing the world.[14] This tends to imply a linear model of the relation between science and society, whereby pure science provides facts that then become available for later policy making. Scholars often portray this model with a riverine metaphor: science is upstream and comes before downstream politics. A fundamental premise of this book, however, is that with regard to questions of sustainability, if not more broadly, this linear model is inadequate. Scientists are now called on to do more than just provide information;

their interaction with society must become paramount. That is, public involvement in science needs to move into earlier phases of the research process, including the selection of metaphors.

In promoting such a view, I invoke all those scholars in the social sciences and humanities, as well as many natural scientists, who have recently critiqued the idea of a pure science that comes before policy. This authoritarian, top-down view of science, separate from society, has been flatly rejected as the "first wave" of thinking about their relation. Science alone cannot decide how we want to live. Instead, we need to recognize the contributions of science as a way of knowing, while simultaneously placing it in the context of more democratic decision making. Such democracy is particularly necessary for environmental and other sciences whose results tend to be uncertain from the perspective of both facts and values and highly contested in the realm of policy, and which must therefore include broader forms of stakeholder-based quality control than normal science. This is the realm of what Silvio Funtowicz and Jerome Ravetz called postnormal science.[15] To a great extent, this book represents an investigation into how we might think about metaphors in a postnormal science of the future.

To address the relation between metaphor and sustainability, we must keep the key concept of polysemy foremost in our minds. Metaphors continue to draw on all the concepts and values from their sources, which have been called "associated commonplaces." Colloquially, I refer to these connotations and value associations of a given metaphor as *metaphoric resonance*, and to the connections this creates among different cultural realms as a *metaphoric web*.[16] Terms such as *exotic* and *invasive*, prevalent in our thinking about species introduced from one place to another, for example, are associated with foreign policy, so it is perhaps to be expected that some critics consider biologists' approach to them xenophobic or even war-mongering. Elsewhere, a similar resonance has incited biologists to argue that certain metaphors in environmental science, such as the slavery metaphor in entomology, are racist or misogynistic. Over the past few decades, as a final example, ecologists have recognized the benefits of disturbance for ecological functioning. Unfortunately, it has sometimes been difficult to convince policy makers of this insight, not only because of the prevalent view that nature is balanced, but also because disturbance has such negative popular connotations. Given what is at stake, we need to understand rather than deny the capacity for such metaphors to conflate technical and ordinary meanings. This conflation is an unavoidable risk of communication, especially given the differing requirements of communication in different realms. Just as biologists need to study a species' evolutionary ecology—its ecological relations within an evolutionary context—to conserve it, by examining metaphoric resonance we will better understand the evolutionary ecology of metaphor to influence such conservation.[17]

Metaphoric resonance warrants a revised view of science communication. Traditionally, scientists have understood the purpose of their communication to be remedying the public's knowledge deficit. Extensive social research, however, suggests that this deficit model is inadequate because it ignores how we now live in a knowledge society where information is overabundant, and where we depend on experts in diverse realms to obtain it.[18] Scientists from one field may be just as ill-informed in other fields as nonscientists, so we are all susceptible to a knowledge-ignorance paradox, where the rapid increase in overall knowledge means that we must be increasingly ignorant of developments in many fields. Specialization has led to

prohibitive entry costs and speech barriers to understanding
and communicating information, so people have little incentive
to learn arcane details of a particular field unless it is directly
relevant to them.

The deficit model stands in the way of postnormal science because it reinforces a hierarchy between scientists and
society. It implies that science is too difficult for most people
to understand, and thus that scientists can communicate only
by simplistic popularization. Sometimes scientists do simplify
to communicate and educate, but as the sociologist Stephen
Hilgartner has demonstrated, popularization "is a matter of degree" because there is no distinct line between real and popularized science. The conception that such a line exists elevates scientists into purveyors of objective knowledge, knowledge that
everyone needs, and nonscientists become dependent on how it
is presented to them. This hierarchy gives tremendous epistemic
authority to scientists, not least because objective knowledge
is thought to have so much currency in the political realm. Yet
many people have rebelled against this structure to ensure that
scientific research addresses their everyday priorities.[19]

These reflections are not meant to deny problems with
science education around the world, but to recognize that their
existence may say just as much about scientists' communication as public credulity. A focus on the latter reflects the conduit metaphor for human communication, another metaphor
like Time Is Money that significantly influences our lives.[20] In
brief, how we speak about communication implies that it is
a matter of transmission. Communication involves putting
ideas and emotions, as "objects," into our words and expressions, as "containers," and transmitting them along a conduit
to another person who understands by removing the objects
from the containers. Applied to science communication, this

model entails scientists expressing facts with their spoken or printed words, facts that nonscientists need to retrieve from these words to obtain understanding. The danger is that this model implies that scientists simply need to express information, and recipients can interpret it clearly and with little effort. In actuality, this is unduly optimistic, because other people must construct meaning from our words and expressions on the basis of their own knowledge, perspective, and values. It is thus difficult to communicate accurately, yet the conduit model defers responsibility by downplaying the role of interpretation. Communication must instead be understood as a process based on negotiation between all parties, and in recognizing this we develop a more realistic view of science communication. In a functioning democracy, not only do we allow such interactive communication, but we also encourage or demand it.

Consider my experience on a bus in Victoria, British Columbia, a few years ago. During an unseasonable snowstorm, I overheard a woman in the seat ahead of me state that it was evidence against global warming. At one level, this clearly reveals a lack of understanding of climate change, yet at another level, this interpretation is totally reasonable. When scientific concepts are expressed in daily language, it is understandable that people will interpret them in the context of their lives. As the MIT historian and philosopher of science Evelyn Fox Keller explained: "The use of a term with established colloquial meaning in a technical context permits the simultaneous transfer and denial of its colloquial connotations. . . . The colloquial connotations lead plausibly to one set of inferences and close off others, while the technical meaning stands ready to disclaim responsibility if challenged."[21] With this model in mind, scientists can attribute misunderstanding to the public, but in the interests of a postnormal science, communication requires effort by both parties. Scientists are responsible for their metaphoric choices, and citizens are responsible for learning to interpret scientific metaphors.

Studies of cognitive frames provide a further insight for science communication. Frames are cognitive structures that "organize central ideas, defining a controversy to resonate with core values and assumptions." For example, Republicans in the United States have adopted frames of "scientific uncertainty" or "unfair economic burden" for climate change, whereas Democrats might frame climate change as a "Pandora's box" or a matter of "religious morality."[22] These frames highlight for people what is significant, metaphors being one of the specific devices that express a particular frame. The point is that facts alone do not communicate. People are not passive receptacles for facts, as the conduit metaphor implies, but instead bring their own perspective to bear on information presented to them. And it turns out that frames trump the facts. Political and religious views better predict concern about climate change, for example, than scientific knowledge. To reach people and to motivate them to act, one needs to appeal to their underlying values, so that the message is meaningful to them. If one cannot place information in the context of their frame, then they will not accept a case, no matter how compelling it might seem. The challenge for science communication is to be true to the science but to appeal simultaneously to people's values.

Reforming Environmental Metaphors

To address whether the metaphors of environmental science promote the sustainability outcomes we seek, I highlight their social dimensions.[23] We need to go beyond the question of whether they are apt descriptors, which is the prevailing theme

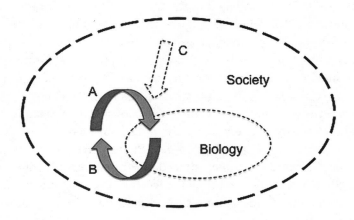

Figure 2. The circular movement of feedback metaphors between biology and society. When biologists select a metaphor (A), they may endorse particular cultural values and assumptions, which may reinforce them within our thought, language, and worldview (B). A key question addressed here is whether alternative metaphors (C) might break this cycle and be more consistent with a sustainable future. This model is quite linear, unlike the web metaphor introduced in the next chapter.

in the literature on scientific metaphors, and carefully consider their wider implications. We seek not just ecological sustainability, but more encompassing socioecological sustainability. We want a sustainable relationship between humans and the natural world rather than sustained ecological systems without humans, which, to many of us, would be a sign of failure. Thus, rather than emphasizing the fertility of metaphors for further research, I promote a broader conception: are the metaphors we choose fertile, or effective, for socioecological sustainability? The plant ecologist Steward Pickett of the Cary Institute for Ecosystem Studies and his colleagues explore such a question, for example, in their interdisciplinary consideration of whether the ecological metaphor of resilience has useful applications for urban design. Such fertility is more germane to the sustainability we really care about. I submit that we cannot reach this objective if we focus only on the epistemic aspects of scientific metaphors and view social benefits and sustainability as mere side effects.

I can now begin to formulate the primary thesis of this book. Environmental metaphors derive from everyday sources, so they reveal to some extent the worldview of the society that coins them. It is not so much that we choose a metaphor; rather, we are chosen by those within our cultural context. Thus, there is a tendency for environmental metaphors to engender a circular feedback between our view of ourselves and our view of nature (Figure 2). Because of the authority of science in modern society, when scientists adopt a social term and give it new scientific meaning, they may simultaneously imbue its associated cultural values with new authenticity. The fundamental question, however, is whether these preexisting social values are consistent with sustainability. To the extent that they aren't, we need to reform our metaphors, even if we have to do so gradually. Without tremendous reflexivity and careful reflection, we merely strengthen the prevailing mind-set by implicitly endorsing it rather than questioning it.

I will argue that many of our environmental metaphors are not the most effective for sustainability, even if they are catchy. Indeed, it may be that they are too strongly rooted in the cultural context of our habits. To make the shift to sustainability, we must overcome embedded faults in our socioecological systems. If the dominant discourse is unsustainable, we need to address it and select an "outlaw discourse." We require new memes to facilitate cultural evolution, and metaphors may play a critical role here as forms of "social disturbance." The linguist Andrew Goatly aptly communicated these ideas in the title of his book *Washing the Brain:* our unsustainable habits are to a significant extent tied to unconscious metaphorical ideology with which we have been brainwashed.[24] The solution is to wash our brains to see afresh in a manner more consistent with sustainability. The present volume is to some extent a cautionary tale, connecting tales of past metaphors with what might happen in the future, given our current and future metaphoric repertoire.

Similarly, Lakoff's book *Don't Think of an Elephant!* argued that American conservatives have grown stronger over the past few decades in part through careful management of their language.[25] Conservative think tanks have produced training manuals for young members that standardize their lingo in the interest of swaying listeners, consequently creating a degree of unity that liberals lack. Liberals are diverse and have no common language in which to express their values, so they rely on available frames, which often derive from effective conservative language planning. In response, Lakoff provided suggestions on how liberals might communicate their values more consistently and effectively.

In seeking linguistic reform, this project aligns as much with deep ecology as with regular academic ecology. Arne Naess, a Norwegian philosopher and one of the founders of deep ecology, summarized this contrast as follows: "The essence of deep ecology is to ask deeper questions. . . . We ask which society, which education, which form of religion is beneficial for all life on the planet as a whole."[26] The metaphor of depth is itself interesting because it implies that "regular" academic

ecology is shallower, which I agree with to the extent that it sets
aside questions of human relationships to the planet. It is in this
spirit that I consider the metaphors of environmental science.
As we will see, the metaphors we adopt today have significant
repercussions for our current and future approaches to environmental problems. Thus, those of us who seek sustainability
cannot ignore how metaphors cross the boundary between science and society; we must become aware of this process and its
implications. Are the metaphors of environmental science as
green as they might be?

This question may raise concerns that I promote the incorporation of values and politics into science. By contrast, I
draw attention to the values that are already there and contend
that such entrenched values may be inconsistent with a sustainable future. Furthermore, many scientists have concluded
that they have a responsibility to advocate for the environment,
even if it means reconsidering the timeless and idealized view
of science as objective and value-free. Others claim that such
advocacy will undermine public trust in scientific objectivity,
although increasingly it is recognized that advocacy is unavoidable. In this debate, however, there has been limited discussion
of language and metaphor specifically. Scientists run the gamut
from condoning hyperbole to seeking a more neutral, objective
language (by "operationalizing" terms with precise technical
meanings, for example), but many scholars conclude that scientific language must be value-laden and must connect with daily
language if it is to affect environmental consciousness.[27] Such
language already occurs so widely in environmental science
that we need to engage these values rather than ignore them.

Numerous scholars have recently attempted to find a middle ground between the extremes of hyperbole and neutrality.[28]
I find Roger Pielke Jr.'s characterization in *The Honest Broker*

to be among the most helpful. Pielke, a scholar at the Center for Science and Technology Policy Research at the University of Colorado at Boulder, reviewed four roles that scientists can play at the policy interface, and he concluded that for many environmental issues, given tremendous uncertainty about ends and means, the role of "honest broker of policy alternatives" is the most appropriate. No amount of science can compel action under these circumstances. Scientists can best contribute by expanding policy options and providing a full analysis of them, including associated uncertainties, so that policy makers can make informed decisions. In some circumstances, scientists may be effective as "pure scientists," simply providing information, or as "science arbiters," as experts sitting on advisory committees who provide input into particular decisions. Pielke demonstrated, however, that scientists in these roles are prone to be "stealth issue advocates," which can lead to a devaluation of science and loss of public trust. Stealth advocacy also encourages activists to wage political battles with cherry-picked scientific data. The problem is not advocacy per se, because Pielke recognized a role for issue advocates in certain contexts where the options need to be narrowed, but advocacy that obscures the distinction between one's science and one's values.

The ecologist Mark Westoby pithily captured the challenges of advocacy when he asked, "Do we have other (ecological) generalizations that are scientifically defensible and also connect to moral and spiritual life?" This is a transdisciplinary question that encompasses the boundary between the two cultures of science and humanities. It seeks claims that are true to science, yet simultaneously speak to broader values. When scientists debate how to walk this line between objective description and cultural salience, they are weighing metaphors. In this domain, they would be wise to turn to poets, "on whose shoulders the future rests." The ethnobiologist and poet Gary Paul Nabhan, for example, praised the value of cross-pollination between practitioners of science and poetry, because otherwise "their isolated endeavors will atrophy, wither, or fall short of their aspirations." He showed how poetic insight helped inform ironwood conservation in deserts of the American Southwest by reframing these trees "not as ugly ducklings but as swans elegantly adapted to the sea of desert surrounding us."[29] We require an analogous shift with many of our metaphors for the environment, but in doing so we will shake the very foundations of how we usually think about science, and particularly the notion that it can be free from values.

Feedback Metaphors and an Overview of the Book

This book explores the social dimensions of metaphors in environmental science through four case studies of what I call *feedback metaphors*. Feedback metaphor is a neologism for scientific metaphors that harbor social values and circulate back into society to bolster those very values. They interweave science and society in a significant way. They are significant within science because they underlie particular fields of research. In this sense, they are constitutive metaphors, which "constitute, at least for a time, an irreplaceable part of the linguistic machinery of a scientific theory."[30] Even if they appear to be replaceable, they have been widely adopted and structure inquiry along one particular line rather than another.

Simultaneously, feedback metaphors have a wider significance in society because of their prevalence, scale, and cultural resonance. In these senses, they relate to concepts that previous scholars have discussed, such as paradigms and root metaphors. Importantly, feedback metaphors codify predominant cultural values in their value assumptions, which affect the core values of scientific inquiry and thereby dictate how research proceeds within a field.[31] Feedback metaphors function not only within science; they resonate with widely held cultural values, thereby leading to the circularity discussed above (Figure 2). They form part of a grand cultural meta-narrative. Metaphors of this magnitude have tremendous significance for how we live our lives and relate to our world.

In the interest of reflexivity, that is, applying my analysis of scientific metaphors to my own metaphors, I considered a variety of names for feedback metaphors. On the basis of discussions with colleagues, some of them, such as *blended, chimeric, circular, global, horizontal, hybrid, moral, nomadic, palimpsest,* and *universal,* seemed to have a confusing resonance or misleading connotations. I could alternatively refer to the process discussed herein as a metaphoric meltdown, but this hyperbole seemed inappropriate (see chapter 6). Other options, such as *socio-scientific* and *value-laden constitutive,* seemed too dry and analytical. I chose feedback to communicate the self-reinforcing and self-fulfilling character of these metaphors and the challenge of breaking the cycle with new metaphors. Those that are currently ingrained may engender positive feedback, yet novel ones that we introduce may be a form of negative feedback contributing to new ways of seeing and relating to the world. As Patricia Ann Fleming, a philosopher at Saint Mary's College, Notre Dame, has explained, we require "moral metaphors" that help create a virtuous rather than a vicious circularity in the interlinking of the human and natural realms.[32]

The historian Nancy Stepan provided a classic example of the operation of a feedback metaphor in her explanation of the mid-nineteenth-century analogy drawn between the "lower races" (as well as apes) and women. The analogy was, first of all, constitutive. Phrenological measurements, in particular, were used to demonstrate a similarity between the size and shape of their skulls, which were thought to correlate with brain volume and thus intelligence. The analogy also encouraged scientists to collect data on the protrusion of women's jaws (prognathism), an observed feature of lower races. Such traits were measured only because of a cultural expectation that women would resemble lower races. Consequently, Stepan observed, "the metaphor . . . permits us to see similarities that the metaphor itself helps constitute."[33] Scientists thus directed their inquiry toward confirmatory data (and unconsciously ignored contrary data), data that seemed neutral and objective only because the analogy meshed invisibly with widespread cultural assumptions. This later allowed a circularity whereby other traits measured in women could become the basis for suppositions about lower races. The analogy also resonated with cultural mores, to the point where it informed thinking about children, criminals, lower classes, and the poor. Stepan concluded that these facts of nature, which scientists observed in lower races, helped reinforce women's place in the social hierarchy.

I examine four feedback metaphors in the following chapters: progress, competition, barcoding, and meltdown. These case studies are not so much meant to be representative as to demonstrate different dimensions of how metaphors operate at the interface between environmental science and society. Yet they are all feedback metaphors from environmental science as defined above, as opposed to less scientific metaphors, such as *ecological footprint,* which originate outside environmental science, or more technical ones that have less resonance with commonly held cultural values, such as *niche.* These boundaries are fluid, however, because such interpretations will vary according to context. Regardless, these metaphors also reinforce respective status quo values: progressive idealism, competitive capitalism, consumerism, and fear-based militarism. These values may not be the most conducive to changing our environmental ethic because they simply buttress forces that prevail in our society and numb people to the need for change. Just as we need to rethink our personal and societal actions in the interest of sustainability, we need to rethink our metaphors.

The first two cases, progress and competition, in chapters 2 and 3, to some extent represent self-fulfilled prophecies because they are biological metaphors coined in the past whose significance for society and sustainability can be seen all around us. Since their adoption in the nineteenth century, they have fully completed the cycle outlined in Figure 2. I draw on the results of a survey to show how they continue to influence society. They are in part significant because they are evolutionary metaphors, which have historically been used to impute human nature, as we interpret it, to nature. This allows people to then state that this human nature is fortified by nature itself.[34] They are powerful, ideological metaphors that justify how we act in relation to the natural world and toward one another. It is thus essential that we rethink them in the interest of long-term socioecological sustainability.

The first case study, on metaphors of progress, elucidates how feedback metaphors interweave science and society in the metaphoric web. Although progressive thought can itself impede the move toward sustainability if we insist on continued growth, people may also adopt it in the interest of sustainability. I demonstrate that there are myriad forms of progress and show how a scientific and a nonscientific group adopt alternative forms to this end. This case study thereby demonstrates how feedback metaphors confound the artificial boundary between science and society, which is one of the prime reasons for thinking more broadly about metaphoric choices in science and scientists' responsibility for them. The focus on progress is also important because a blind faith in the progress of science justifies some people's unwillingness to consider novel forms of science that might be more conducive to sustainability. This chapter also addresses whether one of my own metaphoric choices, that of sustainability progress, should be envisioned as a process or as a destination.

Chapter 3, on competition, investigates a particular way that the metaphoric web is formed, through the entanglement of facts with values. Many scientists have long assumed the distinction between scientific facts and societal values, but this example investigates how metaphors foil this distinction. Survey data reveal how different groups evaluate fact and value dimensions of competitive metaphors in concert with one another, reinforcing the prevailing competitive view of the natural world, rather than a more cooperative one. In the social realm, this competitive bias contributes to the justification of capitalistic values that are often inconsistent with sustainability initiatives. This chapter thereby demonstrates how feedback metaphors can become naturalized, eventually becoming entrenched as normal and true.

Together these two case studies allow us to see the ongoing imprint of past scientific metaphors, which provides an impetus for thinking differently about those we choose now and in the future. Given that scientific metaphors entangle facts and values, we require a revised ethic of environmental metaphor in the interest of sustainability. As a prelude to the final two case studies, chapter 4 offers suggestions for what such a metaphoric ethic might look like and specifically recommends how we might reform the linguistic link between people and the world. I will suggest that we need to do so with respect not only to the relation between humans and nature, but also to our concomitant relation to one another. I turn to indigenous cultures as a rich source of different metaphors that help emphasize such relationality.[35] They embody different empirical ways of knowing the world and alternative solutions to human existence, thereby providing hope that we can, perhaps, shift to a more sustainable path.

The final case studies explore the issues raised in chapter 4 using feedback metaphors coined within environmental science in the past decade or so. Both focus on a particular domain in the search for sustainability, biodiversity conservation. The task that scientists have set out in this realm is undoubtedly challenging, because they encounter a complex domain: millions of species interacting from molecular to global scales over eons. Biodiversity scientists not only want to understand these interactions, but may also simultaneously wish to accentuate the rate and significance of species' losses to engender public concern. A value-laden metaphor such as biodiversity itself may serve both purposes.[36] The case studies show how two prominent scientists have recently coined resonant metaphors that have drawn extensive media attention. As a biologist myself, I will not question their motives, instead assuming that they use these metaphors with the best intention, to make change in the world. Yet that does not mean that their metaphors will be effective in the long run, so my focus concerns whether this is likely to be the case if their metaphors simply adopt the ideology of the day. Furthermore, in both cases the proponents make unjustified claims about the benefits of their projects, which demonstrates a risk of allowing scientists free rein in coining metaphors and evaluating questions that are outside their expertise. Even if my critical perspective is incorrect, it at least demonstrates that the assumptions implicit in such metaphoric choices are open to debate.

Chapter 5 concerns DNA barcoding, a new technological vision for identifying species not by exterior characteristics but by their underlying DNA. The idea is to standardize taxonomy with a single DNA sequence, which might then allow the development of a hand-held device for identification, the Life Barcoder. Proponents of DNA barcoding assume that it will simplify field identification, thus democratizing access to biodiversity and increasing people's appreciation for it. But I ask whether it is likely to nurture relations between humans and other species, given that its value resonance promotes a consumeristic approach to these species. In asking such a question, I also touch on the coincidence of scientific metaphors and technological developments.

Chapter 6 considers how the metaphors of invasion biology, particularly the metaphor of *invasional meltdown,* advocate for native species through fear-based tactics. The assumption here is that fear is needed to activate people's concern, and that it will incite them to take an oppositional, militaristic approach toward invasive species. I review recent research that suggests otherwise. Not only might such language be ineffective in the short term, but it also solidifies a particular hierarchical relation between scientists and society and, in turn, between agencies charged with invasive species control and the general populace. Rather than continued doom-and-gloom scenarios, I suggest that we require socioscientific metaphors that reform human relations in the interest of biodiversity conservation and that harness hope rather than fear.[37]

Together, these final case studies demonstrate that it is inaccurate to view scientists as disinterested because in these cases at least they are using their authority to promote a particular value-laden approach to biodiversity. There is no question that scientists are in the business of promoting values with their metaphors. It is, rather, a matter of which values they should promote, and the extent to which they need to be explicit about them rather than relying on stealthy metaphors. We humans dramatically affect the planet and its denizens through our consumerism and militarism, so these chapters maintain that it is inconsistent to utilize associated metaphors in the interests of reducing our effects.

Chapter 7 brings this discussion to earth by considering whether we might do anything to facilitate the role of feedback metaphors in the interest of sustainability, and if so, how. What metaphoric seeds should we plant for the future? How can metaphors be true to science, but also true to societal values? It might seem that if we open scientific metaphors to poetic, value-based inquiry, we will be less truthful. But if metaphors are resonant, there is no way to avoid discussing them as we would any other claim in the realm of value-laden discourse. Furthermore, if environmental scientists want to continue being social engineers in promoting their work, then they might take the next step toward thinking about what we really want to engineer and how we might do so. We might instead think of scientists in part as poets, for they are not merely embellishing with their metaphors, but extending meaning into new domains.

These questions are too consequential for scientists alone, so we all need to take greater responsibility for metaphoric choices and metaphoric interpretation. Thus, I consider the extent to which everyday citizens might be involved earlier in metaphoric choices, as well as the importance of greater responsibility, and perhaps a new ethic, among scientists themselves. Through this discussion, a core tension in my argument will come to light: even as I assert that we need to choose more appropriate scientific metaphors, I acknowledge that we cannot ultimately control metaphors because of their resonance. The solution to this tension lies first and foremost in greater awareness of the social dimensions of metaphors in environmental science, an awareness that, I hope, this book will help generate. The solution also lies in a vision of engaged citizens and scientists collectively generating a plurality of metaphors, and scientists acting as honest brokers. This would allow open discussion and debate over alternative metaphors and associated values in the interest of providing a vista of possibility for creating a more sustainable future. But this sounds quite progressive, so it is to the metaphor of progress that I now turn.

II
Progress
A Web of Science and Society

Darwinism is social because science is. And of all science the theory that links humanity to the history of nature is likely to be most so.
—Robert Young, "Darwinism Is Social"

If we were not so single-minded
about keeping our lives moving,
and for once could do nothing,
perhaps a huge silence
might interrupt this sadness
of never understanding ourselves
and of threatening ourselves with death.
Perhaps the earth can teach us

as when everything seems dead
and later proves to be alive.
—*Pablo Neruda, "Keeping Quiet"*

The metaphor of progress weaves a tangled web of science and society. One scholar has defined progress as the idea that "mankind has advanced in the past— from some aboriginal condition of primitiveness, barbarism, or even nullity—is now advancing, and will continue to advance through the foreseeable future."[1] This denotative definition places the emphasis on human *cultural* progress, but progress has greater significance as one of the grand metaphysical meta-narratives in contemporary Western societies. As such, it takes on multiple guises that apply in different realms and give it a mythic quality and hence a large role in the constitution of our worldview. In this chapter, I wish to examine progress as a feedback metaphor, tracing its interconnection with biology, its transformation over time, and its relevance for thinking about socioecological sustainability.

People lived and breathed progress in Victorian England. Although progress was implicit in the ancient idea of the Great Chain of Being, it came into its heyday in the eighteenth and nineteenth centuries because of the continued technological success of science. In accord with the definition of cultural progress, British citizens saw themselves as the vanguard in the long march of human history. Furthermore, once Charles Darwin showed that humans had evolved from other species, it became quite intuitive to think of cultural progress as simply the cutting edge of earlier *biological* progress. The metaphor of progress allowed society to personify nature in its own image, confirming the idea that we were the apotheosis of creation and that the way humans *were* was the way evolution had been aiming. The metaphor thereby united the social and natural worlds, as it was applied to the natural world and then back onto humanity (Figure 2, chapter 1).

Evolutionary biologists were not immune to such metaphoric transference. In his magnum opus, *Monad to Man: The Concept of Progress in Evolutionary Biology,* the philosopher Michael Ruse concluded, "Evolutionary thought is the child of [cultural] Progress."[2] He demonstrated that biologists driven by progress-oriented political agendas searched for evidence of biological progress to confirm that cultural progress is natural. This trend was particularly noticeable among popularizers of evolution such as the English philosopher Herbert Spencer, whose writings entrenched progress within the public notion of evolution. But progress also became a constitutive metaphor within evolutionary biology itself, dramatically influencing the theories of such leading evolutionary biologists as Ronald A. Fisher, an Englishman, and Sewall Wright, an American, into the 1930s and 1940s. The distinction between progress as an empirical biological question and as a social perception became increasingly blurred.

Later in the twentieth century, however, the era of progress began to wane. Although many critics had earlier questioned the narrative of progress, including some who sought to counter what they saw as widespread social degeneration, it became increasingly difficult to view human history in a progressive light because the dark side of scientific and technological development became apparent: world wars and environmental degradation. Elsewhere, eugenicists' attempts to spawn cultural progress contributed to the horrors of Nazism. In the late 1960s, the American evolutionary biologist George C. Williams wrote an influential book that lambasted biological progress because it violated epistemic norms, and also because no matter how progress was defined, humans always came out on top—an unacceptably blatant personification.[3] Its rejection was not just epistemically based, but rooted in a broader cultural context. Nonetheless, we have limited knowledge of the current status of biological progress, though as we will soon see, progressive biological metaphors live on in one form or another.

Contrasting Views of Biological Progress

To obtain insight into the current status of progressive metaphors and how they interweave science and society, I conducted an Internet-based survey of members of four organizations. Respondents were asked, among other things, to evaluate eighteen statements containing competitive and progressive evolutionary metaphors. There were 1,892 usable responses. Here I focus on the contrasting response of members of two organizations, the Society for the Study of Evolution (SSE) and the Foundation for Conscious Evolution (FCE), whose members I will refer to, respectively, as the evolutionary biologists and the evolutionaries.[4] For simplicity's sake, I will not discuss results for two other organizations, the Human Behavior and Evolution Society and the National Association of Biology Teachers, until the next chapter. Although I would have liked to survey members of other organizations, such as the Ecological Society of America, their policies did not allow the release of membership information essential for survey participation.

I will focus on their response to a single, representative

statement, "Progress typifies the evolution of life on earth." Respondents were asked to evaluate this statement in terms of the following pair of questions: "Do you believe [this statement] to be factually true? In your opinion, has biological research provided sufficient evidence to support it?" With these questions I sought their view of the veracity of progressive evolution, although it is important to recognize that the statement operated at the level of a gestalt because it asked respondents to characterize evolution as a whole, generalizing beyond specific empirical cases as well as the exceptions that always operate in biology. Despite this limitation, the results discussed here were supported by a factor analysis of the other progress-related statements in the survey.[5]

The survey found that evolutionary biologists agreed with this statement much less than the evolutionaries: 46 percent of the former strongly disagreed with it, compared to only 8 percent of the latter (Figure 3).[6] To interpret this result, we can begin by looking at results for the former organization. The SSE is the world's largest organization of evolutionary scientists, comprising 2,900 members in fifty countries who promote "the study of organic evolution and the integration of the various fields of science concerned with evolution."[7] Its members claim a long lineage to Darwin himself; many refer to themselves as neo-Darwinians. It is hence unsurprising that they had more formal education about evolutionary biology than the evolutionaries, which would certainly have influenced their response to questions about biological progress.

Evolutionary biologists' rejection of the notion of biological progress relates to their belief in *scientific* progress. Ruse demonstrated that biological progress became less and less popular among evolutionary biologists through the past century in part because of doubt about cultural progress, but

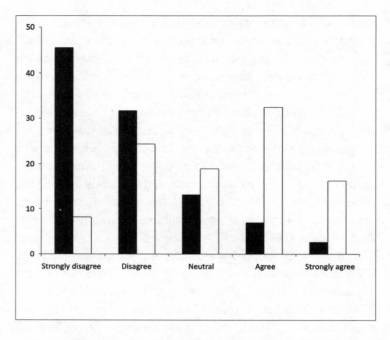

Figure 3. Percentage of the evolutionary biologists (dark bars) and evolutionaries (light bars) who disagreed or agreed with the survey statement "Progress typifies the evolution of life on earth."

predominantly because they had to exclude the cultural values of biological progress to professionalize their field into a valid scientific discipline. To progress scientifically, they had to reject biological progress. Additional results from the survey support this contention. Three-quarters of the evolutionary biologists strongly agreed with the statement "Our understanding of evolution is better than it was 100 yrs ago."[8] Although this might seem commonsensical, it nonetheless contains a value judgment that will be discussed below.

Even if evolutionary biologists have rejected progress,

however, that does not mean that their worldview is any less metaphorical. As Robert Richards, a University of Chicago historian and philosopher of biology, elegantly stated, "Neo-Darwinians seem to have reached general agreement that [the proposal] that evolution is progressive . . . should be dismissed. These ideas, nonetheless, were the very seeds of Darwin's thought. And if one pries back the husks of rhetoric, these earlier stages in the formation of evolutionary theory can, with a little imagination, be yet discovered forming the fetal structures of the views of many contemporary neo-Darwinians." Whereas only 9.5 percent of them agreed or strongly agreed with the statement about progress, for example, fully 50.5 percent of them did subscribe to a statement that simply replaced "progress" with "increasing complexity": "Increasing complexity typifies the evolution of life on earth." Some biologists have touted complexity, another metaphor, as an objective measure of progress, but it is notoriously difficult to quantify, and any increase is neither uniform nor universal because many lineages do not progress at all—or at least not most of the time. An increase in complexity over time could also result from what is known as a random walk: if organisms were very simple to begin with, then evolution can only increase their complexity. Even the desire to quantify complexity could derive from an interest in confirming that *something* progresses; hence, the Duke University evolutionary biologist Daniel McShea wondered, in his empirical review in the pages of the journal of the SSE, *Evolution,* whether "the word 'complexity' (as it is commonly used) is just a modern substitute, a kind of code word for perfection, progress, and proximity to us."[9]

In contrast to the evolutionary biologists, most evolutionaries agreed that evolution is progressive. Centered in Santa Barbara, California, this small group passionately seeks to apply evolutionary thought to directing human evolution. Their acceptance of progress may also mirror broader societal views demonstrated by the movie *Evolution* (in which evolution is equated with a growth in complexity), by ads showing humans as the apex of hominid evolution, and by other examples. Rather than rejecting the concept of biological progress, as evolutionary biologists did, it is likely that the evolutionaries accept it for spiritual reasons. They are much more likely to believe in God and to describe themselves as spiritual than the evolutionary biologists.[10] This accords with a worldview that their Web site explains as follows: "Conscious Evolution is a new social/scientific/spiritual meta-discipline. . . . It is a process of giving direction to the evolution of human systems by purposeful action. . . . The promise of Conscious Evolution is nothing less than the emergence of a universal humanity capable of co-evolution with nature and cocreation with Spirit. . . . The ultimate purpose of Conscious Evolution as a world view is to foster the evolution of our species to full potential, based on the harmonious use of all our powers—spiritual, social, and scientific—in harmony with the deeper patterns of nature and the Great Creating Process itself, traditionally called God." Although this worldview may be unfamiliar, it has been adopted by numerous visionaries operating at the interface between evolutionary biology and society, including Herbert Spencer, the English evolutionary biologist Julian Huxley, the Hindu mystic Sri Aurobindo, and the Jesuit paleontologist Pierre Teilhard de Chardin.[11]

The evolutionaries have adopted progress as a crucial plank within their spiritual worldview. Biological progress supports their belief in what I call *spiritual* progress, the idea that individual humans progress morally and spiritually through their lifetimes. Spiritual progress was originally thought to occur through Providence, until humanism ushered in a self-motivated version. This belief derives some support from developmental psychology, which provides evidence for moral maturation through the human lifespan. The evolutionaries demonstrated their belief in such spiritual development by greater agreement than evolutionary biologists with statements such as "Long-term spiritual or religious practices increase one's compassion" and "In general, adults become more spiritual through their lifetime."[12] They have adopted a historical evolutionary narrative consistent with the progressive story on which they base their faith.

Biological progress and spiritual progress are not the only elements of the progressive worldview of the evolutionaries. In particular, we can add another form of progress to the four discussed above, *universal* progress. This form of progress is critical for the evolutionaries because it encompasses the others in a larger framework, forming a progressive cosmic narrative that links the big bang to the origins of galaxies, stars, Earth, life, consciousness, and human culture. Grand cosmologies of this type posit developmental progress for the universe in its entirety. In the words of Eric Chaisson, an American astrophysicist at Tufts University, "What Darwinism does for plants and animals, cosmic evolution aspires to do for all things." Accordingly, universal progress allows the evolutionaries to incorporate the purpose of their lives, which is connected to spiritual progress, with the unfolding of the universe as a whole. Biological progress, spiritual progress, and to some extent scientific progress are elements of this unfolding, an unfolding that was envisioned by both the mystics and the scientists mentioned above who sought an encompassing worldview. Three current expressions of such thought are cosmic evolutionism, the Epic of Evolution project, and the FCE. These narratives intend to provide humans with a scientifically valid creation story, "Everybody's

Story," which can unite us as we design a new future.[13]

The survey detected progressiveness elsewhere in the evolutionaries' worldview. I expected that they would reject the statement "Long-term evolutionary change is often caused by random drift," because of the antiprogressive resonance of *drift*. The survey found that they often attributed evolutionary change to intelligent design and purpose, so the implications of drift would be inconsistent with their belief system. The results supported this prediction: the evolutionaries were nearly twice as likely to reject this statement as the evolutionary biologists.[14] It is unlikely that the evolutionaries understand the technical meaning of evolutionary drift, but even if they did, they would probably still reject it. Until a few decades ago, evolutionary biologists also rejected drift as an important factor in evolution.

The Metaphoric Web of Progress

We have now seen that biological progress has been rejected by one group in the interests of scientific progress, even as another group accepts it in the interests of spiritual progress. Such is the power of polysemy. There is Progress, and then there is progress, of which there are diverse forms that together constitute a metaphoric cluster founded on polysemy. These forms of progress highlight its metaphorical function, its ability to reach diverse domains of human experience. Although it might seem that the evolutionaries represent a fringe group that simply does not have the facts straight, they instead demonstrate how scientific metaphors are actually adopted and used in social contexts. We cannot pretend that they can be limited to their scientific meanings.

Progressive metaphors demonstrate not only how metaphors refer to one thing in terms of another, but also how they may transfer meanings among discourses. Both senses draw attention to the original Greek meaning of metaphor, "to bear," but the latter more radically expands the usual conception of metaphor.[15] Because of their ordinary roots, scientific metaphors travel between and among scientific disciplines, public policy forums, and disparate social groups. Science is not just some abstract entity; rather, it is conducted by actual people who go home to their spouses, children, neighbors, and newspapers. Conversely, much of the general population is exposed to some science nearly every day. Hence, metaphors move back and forth between these domains on the scale of individual people interacting and talking to one another.

We could use numerous metaphors to conceptualize the role of feedback metaphors like progress in broader science-society interaction. Previous scholars have adopted a "messenger" metaphor,[16] for example, but this suggests a one-way transfer that may too easily invoke the conduit model of communication. It also invokes many of the standard models of metaphor, in which there are discrete source and target domains. Although this model may help us understand metaphors in some contexts, it does not characterize the large-scale, nonlinear feedback among domains in the case of progress or other feedback metaphors.

It is particularly misleading to think in terms of discrete source and target domains when metaphors have entangled cultural and embodied referents. Let us return to progress and consider it at greater length in this context. Progress initially derived from the bodily domain of forward movement, from the Latin *progressus,* which led to its use in the sense of a journey or a march. This did not necessarily imply a goal, but soon the word was applied to a series of events leading toward a better outcome. Progress thus became linked with teleology and the idea of improvement, which is apparent in the senses of progress discussed here. It has more recently extended to describe political views, where progressive is opposed to conservative; in analyzing its use there, the Welsh novelist and critic Raymond Williams noted, "Progressive is a complex word because it depends on the significantly complicated history of the word progress."[17] This complexity accounts for the interweaving of the senses described here, which obviates the need for speaking of discrete source and target domains.

Alternative metaphors or analogies may better communicate this nonlinear conception of metaphoric movement. As a biologist, I considered various biological possibilities. Two of these, *adaptive radiation* and *evolutionary ecology,* seemed too arcane for nonbiologists.[18] Metaphors might also be like bumblebees, transferring value-laden pollen grains among different floral contexts where they may or may not germinate, but this seemed too convoluted.

In the end, I adopted a web metaphor, which entails the diverse interconnections provided by metaphoric transfer (Figure 4). As Harrington explained, "Metaphors do much more than just lend old lexical meanings to new objects: they are literally ways whereby societies 'build' webs of collective meaning; create what I would call *cultural cosmologies* or meaning-worlds that, once built, for better or worse become the 'homes' in which we reason and act, places that constrain without determining any of our particular conclusions or actions." With this in mind, we see that the web metaphor shares some of the benefits of speaking of the "web of life." It highlights context and shows how metaphors embed science within a greater community, a field of contextual meaning, rather then setting it apart. We cannot neatly isolate scientific and popular interpretations of a metaphor because of the nonlinearity and unexpectedness of

Figure 4. The web of an orb spider.

metaphoric movement. In a web such as this it quickly becomes impossible to specify the "original" meaning of a term. The web metaphor, finally, emphasizes interdependence, thereby recognizing how singular effects spread more widely, perhaps through the whole of society.[19] Just as a spider can detect the vibrations when an insect becomes stuck elsewhere on its web, ingrained metaphoric values can have repercussions in science and society that may be distant in space and time.

This web metaphor has its shortcomings, too. As a network or systems metaphor, it represents our times, what has been called the "network society." Some scholars argue that such metaphors are overly nonhierarchical; that is, they neglect inequalities and power relations that structure what is there, or in this case the very real relations existing between science and society. Conversely, the web metaphor may suggest greater patterning than actually exists. Rather than envisioning an intricately patterned orb web (as in Figure 4), we must recognize that this web is much more tangled because of the bidirectional movement of metaphors and the chaotic nature of science-society connections.[20] Such webs tend to ensnare or entangle. Scientists and nonscientists may become stuck because of the stickiness of metaphoric values—they may not be apparent from a distance (like a spider web), but once you adopt a metaphor and get close to it, these values begin to take on a life of their own. If we settle on a particular metaphor, we may become trapped by it and our creative options become closed. To the extent possible, I propose that we cultivate a more organized metaphoric web, though we will continually encounter the unpredictability of metaphor (see chapter 7).

The web metaphor nonetheless shows that as metaphors migrate, they facilitate the transfer of ideas, values, and perspectives. This is not unusual but, rather, to be expected. Wherever

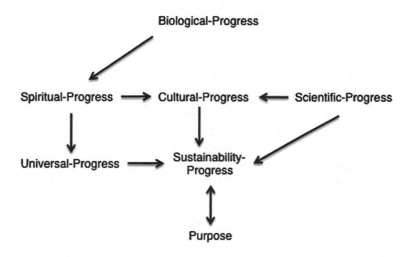

Figure 5. A model of the relation between the different forms of progress adopted by the evolutionaries and the evolutionary biologists; arrows indicate how one form of progress lends support to the existence of another. The evolutionary biologists take the right-hand path to sustainability (via scientific progress, so there is no arrow from biological progress), whereas the evolutionaries take the left-hand side (via spiritual progress). Note the alliance between purpose and sustainability progress. See the text for further details.

they occur, they continue to pull on their associated common-places within the web, which include diverse associations and values in addition to technical meaning. In the lingo of recent sociology of science, metaphors provide a linking component within actor networks. Scientists represent just one component within networks that include the objects they study, institutions that support them, journalists reporting on their work, policy makers, citizens, and others. Across such a diverse array of actors, metaphors promote communication because they act as boundary objects, which are "plastic enough to adapt to local needs and the constraints of the several parties employing them, yet robust enough to maintain a common identity across sites."[21] In this manner, the metaphoric web allows people of diverse backgrounds to engage with and gain some understanding of scientific topics—notwithstanding the challenges this presents.

Progress exemplifies the metaphoric web, the permeable boundary between science and society. Progress is not restricted to the cultural or scientific domain; the five interwoven versions of progress reveal the extent to which this metaphor crosses the boundaries between domains (Figure 5). One could definitely invoke other types of progress, but for our purposes they can be included with those discussed here. For example, technological progress may be subsumed by scientific progress, economic progress by cultural progress, and religious progress, as propounded by, for example, the Templeton Foundation, by spiritual progress.[22]

The metaphoric web might even encompass nonprogressive strains of evolutionary thought based on complexity. Chaisson, whom we met earlier, based his argument for cosmic evolution on a growth in complexity: "Complexity is a, and perhaps *the*, key to both deep and broad understanding in the natural sciences. . . . Material assemblages have become increasingly organized and complex, especially in relatively recent times."[23] Whether or not complexity here is a code word for progress and proximity to us, as McShea suggested, it is being used to think of evolution across scales, from the small, chemical, and biological to the very large, the universe as a whole. This is yet another layer of the metaphoric web of evolutionary thought: progressive and nonprogressive elements, and teleological and nonteleological ones, interweaving with one another to influence thinking across these domains.

These multiple meanings reinforce one another because of their metaphorical interaction. We cannot keep them distinct in practice or in everyday language, regardless of the extent to which biologists reject biological progress within their domain. Even if they operationalize it, they cannot control how it will be interpreted when they release it back into the cultural web in which it originated. Biological progress is always more than just biological progress. It is a feedback metaphor that links diverse aspects of human life, ranging from our personal expectations for our lives through our understanding of the cosmos as a whole. As a consequence, its meaning cannot be limited to any one sphere because biological progress influences so much of how we think about diverse realms of experience: questions about its occurrence are largely metaphysical rather than empirical. It is an elusive shape-shifter that seems to forever imbue our conception of temporal processes. Thus, it dramatically influences our thinking about sustainability.

Routes to Sustainability Progress

The human species currently faces challenges of an unprecedented scale and magnitude. We are changing the planet, and there is widespread agreement that it is not change for the better.[24] Although we may not extinguish ourselves, our actions are leading to a less interesting world and one less supportive of a human presence—let alone the presence of other species. Many people therefore take sustainability as the measure of human activity: does an activity, ongoing or proposed, contribute to

the likelihood that humans and other species will live fulfilled
lives on Earth in the future?

In what follows, I propose to add *sustainability* progress
to the forms of progress discussed above and adopt it as the
critical measure of progress. To put such progress in perspective, Robert Gibson, a Canadian environmental studies scholar,
reviewed how the "old sustainability of customary stability"
that dominated much of human history gave way to the idea of
cultural progress and its eventual idolization over the past few
centuries. Even though its costs then came to light, it was not
rejected, but "instead, the predominant focus was on finding
more viable approaches to progress. The objective was to replace
short sighted and merely economic growth with development
options that promised more comprehensive and lasting gains.
This was the context in which sustainability re-emerged. Unlike
the old sustainability, the new version had to be constructed
not in a world of tradition and preservation of the tried and
true, but in a world of change and devotion to improvement."[25]
Although the Voluntary Human Extinction Movement might
disagree, I assume that most people, in cultures around the
world, could adopt such a vision because they wish human being to continue. It is perhaps a progress that we can all aspire
to because it envisions a future in which humans continue to
exist and adapt, mutually, with other life on Earth. We may
reach such sustainability in various ways, and there is always
the risk of succumbing to utopian dreams, yet I will accept it
as a reasonable objective for our species.

It might seem ethnocentric to emphasize this notion of
sustainability progress. It is certainly still progressive in thinking that there can be a sustainable relation between humans
and the planet. I can state only that this is hope in the face of
potential nihilism and despair. As the sociologist Robert Nis-

bet concluded in his history of the idea of progress, it has on the whole "done more good, . . . led to more creativeness in more spheres, and given more strength to human hope and to individual desire for improvement than any other single idea in Western history."[26] The trick in the current era is to harness such thinking to sustainability.

To aid thinking about sustainability progress, consider the tale of a man who runs out of an airport and hails a cab. He jumps into the cab and yells to the driver, "Hurry up, I'm late." The cabbie tears off through the traffic. A few minutes later, the passenger inquires, "Where are you taking me?" to which the cabbie replies, "I don't know, but I'm going as fast as I can." I suggest that environmental scientists often behave something like the passenger in this vignette—the taxi being environmental science itself. We're going as fast as we can, rarely stopping to consider where to go or whether we have found the right means to get there. We tend to focus on the means for getting "there," giving little reflection to what that "there" is or what the purpose of the striving is. Sustainability thus transmogrifies into sustainable development, allowing the implication of "sustained" development to prevail, despite inadequate focus on lessening it to levels that can truly be sustained.

We can see a parallel issue with the evolutionary biologists and evolutionaries. Both seek sustainability progress in the broad sense just defined, but the specific sense of progress they invoke varies (Figure 5). The evolutionary biologists rely on scientific progress, whereas the evolutionaries commit to spiritual progress, beliefs supported by other portions of the metaphoric web. While this web is alluring, there are liabilities of scientific progress and spiritual progress as routes to sustainability in themselves. In other words, are they the right taxis?

To consider this question, it may help to draw an analogy

with biological progress. We can first clarify biological progress further by defining it as "gradual directional change embodying improvement" through evolutionary history.[27] Biologists might find the first element of this definition—gradual directional change—to be empirically unclear, yet they would find it less controversial than the second element—concerning improvement—because gradual change is intrinsic to neo-Darwinian naturalism. We can contrast these two elements with a directional statement that organisms are becoming pinker over time versus the claim that this is progressive improvement because the pinker organisms are somehow better. The latter claim would be more contentious among scientists because of its valuation component.

To develop our analogy, we can then distinguish relative from absolute biological improvement.[28] Relative improvement accrues through gradual change against a standard within a lineage over time, which is most evident in directional trends in the fossil record that result from the action of natural selection. Natural selection adapts organisms to the environment in which they occur, and their "fit" may become better over time, representing an improvement. As a simple example, fish are far more at home in the ocean than we are. Nonetheless, a problem arises when we attempt to extend thinking about such microevolutionary progressive processes to a macroevolutionary scale, the realm of absolute progress.

With absolute progress, life improves on the whole even among evolutionary lineages. Absolute progress conforms to the traditional notion of a Great Chain of Being: later organisms scale a ladder toward God. Early biologists invoked this view when they drew trees with a ladderlike main trunk leading to us, unlike more recent imagery, such as that of a bush, which represents the contingency of evolution.[29] This shift recognizes the challenge of providing an unbiased measure of absolute compared to relative progress. What is our absolute standard of progress? Will it be size, complexity, intelligence, DNA content, or some other feature? How will we measure these features? How will we weigh them against one another? Consider that parasite lineages often become morphologically simpler over time, though they simultaneously become more specialized in adaptation to their niche. We might justify using either morphology or specialization, in this example, as evidence of relative progress, but how would we assess ultimate progress? This is a particular challenge, given our propensity for measures that place humans at the apex of evolution.

So how does this apply to the evolutionaries and the evolutionary biologists? I propose that they are taking their own relative form of progress and mistaking it for a more encompassing sustainability progress. Of course, this is not their only objective, but, as outlined above, this axis must be primary in our minds. Consider the evolutionaries' belief in spiritual progress. It plays a key role in their view of sustainability because the evolution of human consciousness will allow us to solve environmental problems by directing human evolution in a more sustainable direction. This is expressed in their purpose statement: "Those of us alive today are the first generation born with the choice of learning conscious, ethical evolution, or suffering devolution and the destruction of our life-support systems. . . . We are the generation of the gap between 'here'—a highly technological, over-populating, polluting, brilliant species on the brink of social and environmental chaos, and 'there'—a compassionate, spiritually awakened humanity, capable of transcending the limits of our past human existence and moving toward the unknown."[30] This vision emphasizes the role of spiritual progress as a means toward sustainability progress.

The evolutionaries' adoption of sustainability progress faces a couple of obstacles. Along with the other visionaries mentioned above, such as Chaisson, they tend to validate universal progress with reference to science, biological progress being one of its main elements. As we've seen, however, their views on biological progress contrast markedly with those of evolutionary biologists. It's not that they need to correspond with science if they are to be valid, but rather that if they don't, the evolutionaries may have to rein in the universality of their vision. More to the point, would such spiritual progress lead to sustainability progress? The risk is that spiritual progress can give priority to the human species and thereby isolate us from the natural world. In the words of Gould, "This fallacious equation of organic evolution with progress continues to have unfortunate consequences. Historically, it engendered the abuses of Social Darwinism. . . . Today, it remains a primary component of our global arrogance, our belief in dominion over, rather than fellowship with, more than a million other species that inhabit our planet." One-third of the evolutionaries, for example, agreed with the statement "Humans are the apex of an unfolding universe."[31] It might have thus seemed consistent for one of them whom I met to advocate spiritual progress yet nonetheless order the killing of a matriarch rattlesnake. Unfortunately, if we humans treat the Earth as passive support for our individual or collective spiritual advance, then this advance may ultimately fail. Relative spiritual advance is not the only route to absolute progress or, put more pragmatically, for humans to develop more sustainable ways of living on the planet.

Let us now turn to the evolutionary biologists, on whom I will focus, given their greater pertinence to the argument of this book. Although the evolutionary biologists reject biological progress, this is partly to support an implicit belief in scientific progress. We have already seen that they feel that evolutionary knowledge has become better over time. Furthermore, nearly 75 percent of them also strongly agreed with the statement "Evolution is a fact." Taken together, these results demonstrate a strong belief in the solidity of evolutionary knowledge. Evolutionary biologists may emphasize this to counter creationism, yet it may mean they are being as dogmatic as creationists. As Massimo Pigliucci, an evolutionary biologist and philosopher who is known for his incisive critiques of creationism, observed: "Science is a *method,* not simply a particular body of knowledge. . . . Science, if you will, is always about 'very likely maybes,' never about absolute truth. This is a feature that both creationists and scientific fundamentalists apparently find disturbing and that explains quite a bit of the ideological posturing on both sides of the issue."[32] The term *fact,* by contrast, has a common resonance that lends stability to scientific knowledge, in contrast to a term such as *theory.* It implies not just absolute scientific progress, but the conclusion that we have already arrived at a solid endpoint.

Scholars have long debated, however, whether science progresses at all. Scientific representations certainly seem to become increasingly accurate over time, so within that relative way of framing things science *does* progress. We must nonetheless ask critical questions, such as whether this frame is the right one, whether the particular values we have adopted in our science are actually the best ones, and whether the benefits of science always outweigh its costs. By emphasizing realms in which scientific progress is ambiguous, many writers conclude that it is fictitious. Science in this view does not provide ultimate truth; instead, following Pigliucci, it provides knowledge that is continually susceptible to revision by future observations and empirical testing. Another philosopher of science, Thomas Kuhn, explained this distinction by contrasting

scientific-progress-from, a local progress from what we knew
before, with scientific-progress-to, a progress toward an Aristo-
telian *telos*—truth.[33] Just as biological progress was compelling
and reassuring to evolutionary biologists a century ago, some
of them may still believe in scientific-progress-to, toward the
truth, and this faith may yet trump critique.

Many evolutionary biologists express concern about biodiversity loss and accent the relevance of their field to conservation initiatives. Does regular scientific progress necessarily lead
to this espoused sustainability progress? Like the evolutionaries'
spiritual progress, it has a role to play. The unquestioned assumption that any and all scientific knowledge—and associated technology—contributes to sustainability, however, derives
from faith in the importance of objective knowledge for solving
global problems. Scientists obtain power and become the priests
of our era to the extent that they provide a special form of
knowledge that can be used to do such wonderful things. And
we often consider that the final test of scientific knowledge: we
can *do* things with its results, such as applying it to reverse the
decline of an endangered species. Regardless, we know now
that the linear view of the relation between science and social
outcomes is flawed. Science may allow us to do things, but we
can assess its contribution to sustainability only by incorpo-
rating broader contextual and socioecological questions. We
typically think of sustainability as doing something out there
in the world, when in fact we may need to first reassess the way
we are setting the problem.

If we insist on conceptualizing scientific progress as a linear ascent, perhaps as the growth of a branching tree, I am in-
clined to focus on the interstices, the spaces between the leaves
and the branches, the things we do not now understand and
may never understand. Science does not necessarily display

these interstices. Philosophers of science generally agree, for example, that there isn't One Science; rather, there is a "disunity" of sciences, and our understanding of the world is therefore "dappled." The methods of the physical sciences may not reach these interstices; it may even be dangerous to think that these methods apply in the social realm. Accordingly, many young scientists, having realized that the real problems of conservation are interdisciplinary, have redirected their research in this direction despite its career risks.[34] We need to question the emphasis on the cutting edge of scientific research, a metaphor implying that it is natural for humans to seek answers to the unknown, rather than perhaps recognizing that some things may be ultimately unknowable or that mystery might make life worth living.

More broadly, we require insights into these interstices from the myriad alternative ways of knowing. This is particularly the case in our species' search for sustainability. Sustainability progress will benefit from plurality rather than a focus on (relative) scientific progress. Instead of assuming that Western science is the solution, we need to open our minds to how diverse people have learned to live on our planet. Our objective, together, is sustainability-progress-from rather than progress-to, as Kuhn described them. We need to ponder how different metaphors contribute to sustainability rather than becoming too attached to our own because, like all metaphors, they highlight one aspect and thus miss out on others. In this way, they demonstrate why it is so critical that we look more carefully at the metaphors we choose to adopt and use as a community of people on the path of sustainability.

A Purposeful Life: Personification and Sustainability

It is easy enough to critique visions of spiritual and scientific progress, but at a deeper level they speak to the metaphoric web of progress and its multiple forms. Why is progress, in one form or another, so compelling? It appeals to people in part because it connects with the common human experience of purpose. In the 1998 General Social Survey in the United States, over 90 percent of respondents disagreed or strongly disagreed with the statement "In my opinion, life does not serve any purpose."[35] Embedded in purposeful lives, we are to progress as fish are to water. We have an existential need to ensure that what we are doing matters, that our lives are progressing. This reaches all the way to ultimate questions about the sustainability of the human enterprise.

When we awake each morning, we set out with a purpose—or at least this assumption is associated with a healthy and fulfilled human life. This purpose may be clear, or it may be veiled, but it influences everything we do. It may be as simple as living well and loving our families, or it may mean striving for success or aiming to change the world. We want our incomes to grow, our knowledge and skills to improve, our lives to get better. As scientists, we want our experiments to work, our papers to be published, our careers to unfold, and perhaps our research to contribute to environmental objectives. We live this way despite contrary evidence—as suggested by the discussions of spiritual and scientific progress above. Scientists continue to have faith that their work will contribute, even if citation indexes show that most papers are rarely (if ever) cited and that citation rates decay exponentially. Regardless of the specifics, we strive to live this way, as opposed to living without purpose, directionless. We want our lives to matter. And this experience of purpose contributes to our sense of progress through the metaphoric web.

Purpose and progress connect with one another through a critical cognitive metaphor in Western culture, A Purposeful Life Is a Journey. If a purposeful life is a journey, then we are all travelers, the objectives we set in life are destinations, and the plans we have for reaching them are our itineraries. Just as we do when we prepare for a real journey, we need to plan carefully to meet our objectives, including ensuring that we have what we need for the journey, such as education, and anticipating problems that we must overcome. We want to make sure that we are on pace and headed in the direction we want, rather than getting lost or being behind schedule. This complex metaphor underlies much of how we approach the world and our lives. Yet it is not universal. We might recollect, as Lakoff and Johnson cautioned, that "there are cultures around the world in which this metaphor does not exist; in those cultures people just live their lives, and the very idea of being without direction or missing the boat, of being held back or getting bogged down in life, would make no sense."[36] This perspective might help us understand how dramatically the metaphoric web of progress affects how we live our lives

Nonetheless, it seems unlikely that we will abandon progressive metaphors anytime soon, but we do have flexibility in giving precedence to one form versus another, whether financial gain, altruistic pursuit of knowledge, or spiritual growth. Yet for any of these human expressions to endure, socioecological sustainability needs to be our fundamental criterion. In particular, in the current context, how might the metaphors we choose within science bring us more in line with sustainability progress? We have to expand the bounds of scientific progress

to include a broader sense of socioecological sustainability. Such
a view of progress may ask deeper questions about how to weigh
scientific advances versus regressions.

Consider in more detail the case of personification in science. Its importance struck me one spring day when my three-
year-old daughter and I were riding our bikes and observing
the lives of creatures around us. I pointed to a pair of crows
sitting near each other on a power line: "Look at that crow
walking over to the other one. What are they doing?" She responded, "Are they going to fight?" That might be an expected
question from a child who plays lots of video games, but we
had no television. It shows how deeply an antagonistic form of
personification has penetrated our culture, including children's
perceptions of the natural world.

Personification is of course a form of metaphor, specifically a metaphor with which we conceptualize something that is
not a person in terms of personhood, such as thinking of genes
as selfish or animals as competitive. Classically, science attempts
to reduce personification, as it taints the object of study with
human attributes. As Ludwig von Bertalanffy, a biologist and a
founder of systems theory, explained, "An essential character-
istic of science [is] that it progressively de-anthropomorphizes,
that is, progressively eliminates those traits which are due to
specifically human experience." The origin of this tendency is
not too difficult to find. Lorraine Daston, a historian and the
director of the Max Planck Institute for the History of Science in
Berlin, has contended that "like us, seventeenth-century natural philosophers condemned anthropomorphism not only as
an intellectual but also as a moral error: failure to recognize
that nature was the other was not simply wrong, but danger-
ously wrong."[37] In this sense it brought Judeo-Christian values
into science rather than removing them. Regardless, it followed

that science had to lessen personification over time if it was to progress.

Some scholars, however, maintain that personification is almost unavoidable in biology. We might dismiss the agency attributed to the Oxford biologist Richard Dawkins's selfish genes, but agency appears to remain even in the proposed alternative conceptions. As another example, evolutionary biologists have removed blatant teleology from their field by using natural selection and other mechanisms to explain evolutionary change. Darwin's writing itself was very anthropomorphic, introducing the idea of natural selection through an analogy with artificial selection, yet even "the newer twentieth-century explications of natural selection that have accompanied the rise of mathematical, experimental, and ecological population genetics have not displaced the older figurative and rhetorical life of the term."[38] For example, who is the selector? Even common conceptions that natural selection "molds" and "shapes" continue this anthropomorphic tendency. We may find these associations with personification disgruntling, but they are unavoidable because we perceive the world from the perspective of embodied human beings.

We thus might expect that evolutionary biologists have a contradictory relationship with purpose. Although they strongly disagreed in the survey that evolution has an aim or purpose, for example, they embrace it in another arena. Consider Pigliucci's statement that "having teeth structured in a certain way makes it easier for a tiger to procure its prey and therefore to survive and reproduce—the only 'goals' of every living being." As the philosopher Ernst Cassirer has pointed out, evolutionary biology thereby elevates purpose, in the form of survival value, to the primary consideration in our study of life: "All other questions retreat into the background before this one." There appears to be an interesting inconsistency here, but even if evolutionary biologists do reject purpose, they cannot base this decision on scientific inquiry, which, having excluded human purpose from the universe through objectification, can only draw the circular conclusion that there is no purpose. Its rejection is an assumption, and one that relies on a firm line between humans and the rest of nature, despite Darwin's dismissal of this metaphysical stance. Scientific evidence lends little input because we cannot even characterize life or consciousness, our most certain experiences, other than as patterns or as what their absence looks like.[39]

Not all evolutionary biologists have rejected the idea that the purpose we experience is something that pervades existence. Consider Wright's monistic panpsychism, the belief that life and consciousness occur to some degree in all matter. He wrote: "Emergence of mind from no mind at all is sheer magic. We conclude that the evolution of mind must have been coextensive with the evolution of the body. Moreover, mind must already have been there when life arose and indeed must be a universal aspect of existence—still assuming that mind cannot arise from nothing." Darwin also contemplated panpsychism in his struggle over the place of consciousness in the natural world.[40] Because we have evolved from the natural world, it is possible that what we experience as purpose is something that all of creation senses (in varied ways).

A similar view may underlie the evolutionaries' universal progress. Whereas I questioned earlier whether spiritual progress necessarily leads to sustainability, one could counter that the evolutionaries are more in touch with a purpose-laden universe. They were much more likely to believe that "nature is spiritual or sacred."[41] In this respect, the evolutionaries appear to believe that the purpose we experience is something that imbues existence.

Anthropologists have shown that many cultures around the world personify the natural world. Frans de Waal, a primatologist at Emory University, took this as an opportunity to remove the "tinted glasses through which [scientists] stare into nature's mirror" by comparing our view with that of other cultures. He reviewed how Western primatologists perceived great apes in accord with their cultural biases: as independent individuals who lacked coherent social groupings. In contrast, Japanese scientists interpreted them through the lens of their society, as part of stable groups, a view that we now take for granted. He quoted a Japanese primatologist, Junichiro Itani, who observed, "Japanese culture does not emphasize the difference between people and animals and so is relatively free from the spell of anti-anthropomorphism." Again, we might have expected evolutionary theory to have dispensed with the idea that humans and animals are qualitatively distinct.[42]

Although we may now recognize the connection between ourselves and the great apes, many cultures extend this much further. For example, the social anthropologist Tim Ingold of the University of Aberdeen demonstrated that the Ojibwa draw a less firm distinction between themselves and other organisms than we normally do: "Persons, in the Ojibwa world, can take a great variety of forms, of which the human is just one." This provides an entrée for a more significant challenge by metaphors on the literal-figurative distinction. As Ingold elsewhere explained: "To posit a 'metaphorical' equivalence between goose and man is not, then, to render 'one kind of thing in terms of another,' as Western—including Western anthropological—convention would have it. A more promising perspective is [to argue] that metaphor should be apprehended as a way of drawing attention to real relational unities rather than of figuratively papering over dualities. Metaphor . . . 'reveals, not the "thisness of a that" but rather that "this *is* that."""[43] That is, perhaps scientists' wish to discredit personification in order to be objective artificially separates them (and us) from the natural world, when we would all benefit from breaking down such distinctions, instead emphasizing our connections with other organisms. Personification might help us understand other organisms as being like ourselves, perhaps helping us develop greater empathy with other species rather than treating them as unequivocal others.

Personification may help us better relate to the world. As Stephen Trombulak, a biologist at Middlebury College, remarked, "That creatures with an innate sense of passion for life should attempt to completely suppress this passion in an attempt to be completely objective is the real lurking inconsistency [in conservation biology]." Perhaps we need to be more passionate and engaged in the interests of conservation. The philosopher Paul Feyerabend generalized this when he claimed that scientists' ideas and actions "can be as destructive as those of their more brutal contemporaries." He noted, "Science certainly is not the only source of reliable ontological information," and then argued, "It was the technological and social success of science that separated purpose and matter, destroyed animism, and turned people into addicts [for its products]." He also wondered whether more of the ancient animism might "create a greater harmony between human enterprises and nature," thereby lessening our environmental impacts. The environmental studies scholar Neil Evernden was even more adamant about the need for environmental thinkers to accept personification, which has been called the pathetic fallacy (though he considers it a fallacy only to the "ego-clencher"), and to stimulate a resurgence of animism.[44] The environment is part of us, so the world is animate. Perhaps a fundamental limitation of modern science, and a crippling factor for developing a more encompassing environmental ethic, is that it delimits us from the natural world.

Personification urges us to revisit how the metaphors we have adopted exist among other possibilities, some of which Western culture might disallow. Some of our metaphors are entrenched and associated with normal science, a paradigmatic way of looking and thinking. Let me offer a prime example: mechanism. We commonly understand life as a mechanism, and biologists often seek mechanistic (and reductionistic) explanations in their research practices.[45] Most scientists, including ecologists, prefer mechanism to personification. Few would deny the benefits of thinking of biological systems—whether ecological or medical—as mechanisms, because it has contributed to many conceptual breakthroughs. However, this way of thinking can also be misleading. The technical metaphor of "ecological assembly," for example, might give the impression that we can figure out how ecosystems are constructed just by knowing more about their parts and how they fit together. This might lead us to devalue existing ecosystems if we think we can replace them with reconstructed versions. Although philosophers and biologists debate the various merits and costs of mechanistic explanations, we may overlook how mechanism too is a metaphor, one that may impede our ability to comprehend life and its processes in other ways.

Furthermore, is it not strange that we are more comfortable with mechanomorphism, understanding the world through machine metaphors, than anthropomorphism? The biologist and futurist Elisabet Sahtouris reasoned, "Mechanics is an invention of just one species, man or *anthropos.* . . . . and so the mechanomorphism required of us was really a kind of secondhand anthropomorphism. . . . Was nature at large not likely to be more like us naturally evolved creatures than like our machines?" The prevalence of machine terminology reflects our modern industrialist world, but the resultant deanimation and desensitization disconnect us from our environment and even ourselves. The Berkeley historian Carolyn Merchant has demonstrated that the mechanistic worldview, led by Francis Bacon's scientific method, contributed to the "death" of nature because the living, feminine Earth had to be subdued to uncover her secrets (the habit of personifying her not having disappeared).[46]

If we assume that the world is mechanistic, we may find supportive results, but that does not mean that the world *is* mechanistic. Yet we may nonetheless begin to see organisms *as* mechanisms. We have seen the brutal implications of such association in how the Nazis conceived Jews as "machine people" on account of their thought and individuality, in contrast to the holistic science and unifying Gestalt of the Germans. In a similar way, the transition in ecology from thinking of communities as organisms to thinking of them as systems of energetic feedback loops gave rise to a "technocratic optimism" in which influential ecologists such as H. T. Odum gave themselves a special role in planning a fully engineered society.[47] If such conceptions cause people to treat other beings like mechanistic machines, then I contend that they are inadequate to the task at hand—and that personifying might be preferable. If our metaphors do not encourage us to maintain a world in which we can live, what is the point of understanding for understanding's sake?

This is not to deny potential problems with personification. For example, the environmental movement has personified whales so that moratoria extend not just to endangered species, but even to more widespread ones. Icelandic whalers, however, have built a whole way of life in small villages around whale harvests. Their view of whales is pragmatic, rather than spiritual or romantic. The harvest is a matter of economic and cultural survival, so the whalers are offended by the fundamentalism of environmentalists and antiwhalers.[48] Although personification can be used to promote an environmental agenda, the downside, in this case at least, might be the loss of a way of life.

We can turn from this discussion of personification to seek a middle ground in thinking about progress and purpose. It may be a mistake to reject progress simply because it is anthropocentric. The English philosopher Mary Midgley has suggested that "purpose-centred thinking is woven into all our serious attempts to understand anything," but "rules have to be worked out again for its proper use."[49] In working out such rules, we do not have carte blanche to accept or reject personification, but we should consider whether sustainability progress might require one form or another in some contexts.

Overall, this chapter has demonstrated the extent to which progressive metaphors interweave science and society. We cannot think of progress in one domain without sending tremors through the web into other domains. And this concluding discussion of personification raises questions about whether environmental scientists have a responsibility to endorse metaphors that are at least consistent with sustainability rather than undermining it. We cannot maintain a faith in scientific progress, which emphasizes its epistemic contributions, without careful attention to the mixed blessings that it often provides. We might be convinced of the value of a particular metaphor, but the pertinent question for us now is whether it is progressive in a broader socioecological context. To address this question, we need to attend more carefully to the place of values in science and its metaphors. We have seen values throughout this discussion of progress, but I delve into them further in the next chapter, on competitive metaphors.

III
Competitive Facts
and Capitalist Values

*The use of metaphoric language in science may allow science
to function as an authoritative voice for "what is" while
simultaneously producing a stream of stories, sotto voce,
on the theme of "what should, could or must be."*
—Anne Harrington, "Metaphoric Connections"

*Central to the world-view in which we currently operate
is the Malthusian metaphor of struggle. Equally, however,
it is clear that other cultures, in other times and places, have
adopted very different views of the world, based on very
different metaphors. It is therefore certainly not beyond the
bounds of possibility that a cultural transition towards sustain-*

able development might proceed, indeed might require us
to proceed, from a different set of ideas, a different world-view,
involving a new cosmology and a different metaphor.
—*Tim Jackson, "Sustainability and the 'Struggle for Existence'"*

P rogress is not the only metaphor that interweaves environmental science and society. Many of our metaphors for nature mirror and reinforce particular social ideals and actions that have profound implications for sustainability. In this chapter, I focus on how we are continually bombarded with images of our basic competitiveness, from our personal urges to the realm of corporate takeovers. Quite often this competitiveness is founded in the natural world: we are competitive because that is the way of nature itself. We may adopt religious or spiritual practices to reduce such base tendencies, but we cannot escape them.

How did we arrive at this worldview? Competition is a metaphor that links science and society, so we can look at this interface as one way to understand how the competitive emphasis came to be. Before Darwin's time, the natural world may have been seen as frightening, but it was not thought of as competitive so much as harmonious. As the German scholar Peter Weingart summarized, "Malthus and Darwin both took part in a period of intellectual change during which the view of nature and society as a harmonious system gave way to one characterized by the 'law of struggle.'" A competitive mindset was frequent in Darwin's culture, as an accompaniment of continued economic growth, so it had an unavoidable influence on biological thought. Although biologists applied the term *competition* to other organisms before Darwin did, his emphasis on it in *On the Origin of Species* helped stimulate its priority among the next few generations of biologists.[1] Ecologists soon considered competition to be one of the fundamental inter- actions between species, an interaction wherein both partici- pants are harmed. A classic example is the competition between individuals for limited resources, whether it is plants competing for nutrients in the soil or predaceous fish for smaller prey.

As time went by, this competitive emphasis in biology fed back onto our self-image. This happened largely because competition became an intrinsic element of thinking about evo- lution itself through an association between natural selection and a competitive struggle for survival. This association arose in part because of interpretive troubles arising from Darwin's term *natural selection*. To counter these problems, Alfred Russel Wallace "persuaded Darwin to use Herbert Spencer's phrase the 'survival of the fittest' as a synonym for natural selection, but one free from any of the metaphorical elements that Wal- lace was convinced had led people to misunderstand Darwin's own term." Ironically, this metaphoric phrase and the related "struggle for survival" proved just as problematic; in particular, they provided one basis for social Darwinism, the application of evolutionary theory within human society.[2]

Many people interpreted Darwinian theory to mean that nature is competitive, and it followed that we should be com- petitive, too. If there is a struggle for survival and only the fittest survive, then people must look at one another as competitors for limited resources. Accordingly, social Darwinism flourished, promoting human actions that conformed with the harsh work- ings of natural selection in the wild. Policies were instituted to lessen the reproduction of the weak and to reduce assistance for the poor, and competitive nature thereby became one of the justifications for capitalism. Having convinced ourselves of our basic competitiveness, our capitalist tendencies—whereby everyone is either a winner or a loser—were given free rein.

With this focus on competing among ourselves, the significance of the natural world retreated into the background. Some argue that the industrial development contributed to a change in our relation with nature from one of cooperation to one of exploitation; it became merely a source of materials for the expansion of capital.[3] The logic here draws on the concept of progress, for if you succeed in the competition within society, you will progress, materially and perhaps spiritually. We move forward with progress, but with competition it becomes a race. The winners in the competition serve as a model for the continued expansion of human capital. Clearly, however, sustainability will be more difficult to achieve if we consider ourselves to be fundamentally competitive and exploitative beings.

Metaphors and Social Values

We have just seen various ways in which competition observed in nature may influence the constitution of society. Many philosophers, however, have a good argument for why society should not be influenced by science in this manner, and they would thus consider this discussion hopelessly confused. Scientific findings are supposed to be separate from social implications because of a distinction between what "is" and what "ought to be," between a factual statement and an ethical one.[4] It is this distinction between facts and values that is the focus of this chapter. This focus is an essential step in the case presented in this book, for it provides a way in which metaphors may become naturalized across the science-society interface. And once we question this fact-value distinction, in parallel with the science-society one, we have justified the consideration of scientific responsibility that appears in chapter 4.

When most people think of science, they probably think of the facts that it provides, its truths about the world, on which we can depend. Yet how do these facts relate to our actions, to what we should do and how we should act? Most of us now accept that the climate is warming, for instance, and largely as a result of human activity. It may thus seem obvious that we should do something to prevent further warming. As philosophers beginning with David Hume have pointed out, however, we cannot logically deduce that conclusion from the initial claim about what is occurring. That is, an "is," a fact about the world, cannot lead us to an "ought," a claim about what we should do. Although it might seem that evidence for climatic warming necessitates that we do something to reduce it, this leap from is to ought requires an additional premise about what we value—in this case, for example, that global warming is a *bad* thing because of its current and potential future effects. Given that people hold different values, the "ought" we might draw from a scientific "is" is not a scientific issue but a moral, social, and political one. Regardless of how universal the need to act to prevent global warming might seem, the political wrangling about enacting policies to do so demonstrates that this is by no means the case.

Let's consider the case of evolutionary psychologists, who attempt to explain current human behavior using our evolutionary history. Critics have observed that they sometimes forestall ethical discussion of their conclusions with the slogan "Ought cannot be derived from is." They might claim, for example, that rape is an evolved and natural feature of human beings historically but then deny that this should apply to present-day policies related to rape. This is misleading, however, because their slogan should be interpreted to mean "Ought cannot be described *exclusively* from is"; that is, factual premises *alone* cannot lead to normative (value-laden) conclusions.[5] But scientific knowledge must be *part of* our moral considerations; it is just not the whole story. Just as this knowledge should be considered in evaluating rape, we must acknowledge evidence for human competitiveness, though that alone does not justify proclamations about how humans ought to be or the institution of policies that encourage this competitiveness.

It turns out that in practice it is difficult to separate is and ought because the implicit distinction between facts and values is not nearly as clear as we might at first imagine. As extremes, for example, we can contrast a more factual statement, such as "The Earth has warmed by 0.13°C per decade over the past fifty years,"[6] with a more normative one, such as "We must prevent continued global warming." By scratching below the surface, we would find that both of these statements are coconstituted by factual and normative elements. At a minimum, the former relies on all the scientific interest in documenting such change, and the latter relies on evidence that there is enough of a problem that we need to do something.

More deeply, diverse studies of science and technology reiterate that there is no such thing as value-free science: facts and values intermingle throughout the scientific process, from its beginning premises, through its practices, and to its policy implications. This is undoubtedly the case in climate change science. Scientists make evaluations not only when deciding what to study, but also in the inner workings of their research, in evaluating the evidence for hypotheses. It is not just bad science that does so, for explicit attention to values can make better science. Physics might be held up as an ideal value-free science, yet it can obtain this status only because "most of physics simply doesn't matter to us."[7] It is not a good model for science that does matter, which includes most contemporary environmental science.

I will return to the question of value-free science later on, but the emphasis here will be on the oughts that people do in fact derive from scientific discussions. A primary reason for this transference is that we express facts with language. And language has its own dynamics. In particular, the meanings of words vary according to their context. Many everyday words, such as *cruel* and *crime,* are thick ethical concepts, which have normative associations.[8] In the environmental realm, for example, *wilderness* is not just a physical noun because it carries various adjectival associations related to how barren, desolate, and threatening it is. As another example, I was once a teaching assistant for a large introductory biology course when the professor told me about a student who had come to his office concerned that the professor's description of female birds as drab relative to male ones was chauvinistic. The professor was shocked at this political correctness: it is a simple fact that male birds are usually more brightly colored than female birds. How could someone possibly take this as a chauvinistic comment? In pondering this story, I have begun to recognize its deeper significance. For unlike Humpty Dumpty, who stated, "When I use a word, it means just what I choose it to mean, neither more nor less," we cannot make words mean exactly what we want them to mean. When we communicate, we will always have to deal with the thickness of particular terms, an issue we will encounter with each of the case studies discussed herein. It is only in the rarefied domain of science that we imagine we can avoid discussion of how facts and values combine.

Scientific metaphors have a particular tendency to resonate with normative associations because they have been taken from a broader cultural context before being applied within science. One of their distinctive features is that they have been used across established boundaries. Even though scientific metaphors may seem disjointed from their cultural source domain, they often retain an implicit (or sometimes explicit) value resonance, which one observer has called "value creep." Although it may be erroneous to naturalize an ought with an is, when we accept a metaphoric is, it may automatically lead to an ought, often implicitly and unconsciously (Figure 6). As Rom Harré and his colleagues explained in *Greenspeak:* "'Empirical observation shows us that the facts are a, b, c, d etc. which we can sum up by saying that we have a case of M (where M is the metaphor). However, anyone who has actually confronted a case of M would know that they *ought* to take a course of action so-and-so. So it seems we should do so-and-so in this case too.'" Stated another way, scientific metaphors entangle facts and values, so what seems simply a scientific fact may actually carry an implicit ought. There are thus policy implications to drawing attention to such metaphors because "the glide from facts to recommendations no longer seems graceful or obvious."[9]

Consider the metaphorical statement "The health of the Great Lakes has recently declined."[10] Some scientists would defend this statement with objective measures of health, such as pollution levels or the population sizes of aquatic indicator species. When we speak about ecosystem health, however, we necessarily draw on both factual and normative components because the choice of which measure or measures to use is value-laden. It depends on how we conceive of health. There has thus been continued debate about whether ecosystem health can even be considered a scientific concept. Those

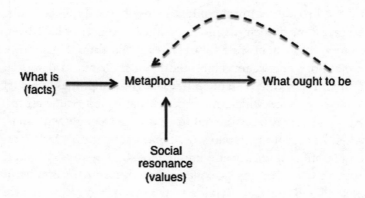

Figure 6. The process by which scientific metaphors may convert
what "is," as determined by scientific investigation, into what "ought
to be." The metaphor summarizes the relevant empirical evidence
but at the same time may carry a social resonance. Further note
that an implicit ideal of what ought to be may itself influence
a scientist's choice of a metaphor (as indicated by the dashed line).

scientists who use it as such entangle the facts with the popular
resonance of the word *health*. Most of us would interpret a
statement about ecosystem health to be value-laden because
we consider health to be a positive feature within our normal
personal experience. A factual statement thereby easily becomes
an evaluative ought, namely that we should seek to return the
Great Lakes to a state of renewed health. A similar process applies to many terms in environmental science, most notably
biodiversity, which assumes that diversity is a good thing. I do

not wish to claim that such statements are straightforwardly problematic; instead, this type of terminology may exemplify the sort of resonant language that environmental scientists need to cultivate, though carefully.

Health is just one example of an environmental metaphor that blends facts and values, thereby circumventing the hallowed distinction between them. Such metaphors naturalize particular values because it is their foundation in what *is* that gives their normative dimensions such potential power. They are not just statements of fact, but facts that prescribe particular values (often without seeming to do so). Philosophers and logicians might call this a fallacy of equivocation, a form of invalid argument where the meaning of a word shifts in mid-argument,[11] yet it is a normal process with metaphors. Much of our communication is polysemous, metaphors in particular having different meanings depending on circumstance, meanings that disregard a tidy distinction between fact and value. Let us now turn to the example of competition, a feedback metaphor that has been with us for such a long time that it has become naturalized and thus structures our understanding of the world in which we live.

Competition and Our View of the World

The example of competition will allow us to see in more detail how metaphors naturalize particular oughts by entangling facts and values. As I discussed above, it was our perception of competition in the cultural world that contributed to a large extent to our search for it in the natural world. Having found it there, it became the way things are. Once the metaphor was naturalized in this way, people could more easily defend it in the cultural realm: not only is competition found in societies, but we should actively promote it because it is the way the world works—it is natural. As the environmental sociologist Michael Carolan put it, "Placing something within the realm of the natural relinquishes our ability to critique it."[12] The metaphor therefore becomes invisible, and we eventually reach a point where a metaphor such as competition is effectively dead because we hardly recollect its metaphorical roots. In this manner, a metaphorical worldview becomes stabilized through the merging of values with facts.

To what extent has competition been naturalized as a metaphorical worldview? To address this question, I again consider results from the survey introduced in the previous chapter because it also asked respondents to evaluate statements about competition and cooperation. I found that over three-quarters of the individuals I surveyed agreed that the statement "A struggle for survival characterizes evolution" is empirically supported (Figure 7). Not only did about 70 percent of the evolutionaries and evolutionary biologists agree with this statement, but so did members of two other organizations, the National Association of Biology Teachers (NABT) and the Human Behavior and Evolution Society (HBES), which are hereafter referred to as biology teachers and evolutionary psychologists, respectively. In contrast to evolutionary biologists, who emphasize the study of nonhuman organisms, evolutionary psychologists focus on human behavior, as mentioned above. As it states on its Web site, the thousand-member HBES is "an interdisciplinary, international society of researchers, primarily from the social and biological sciences, who use modern evolutionary theory to help to discover human nature—including evolved emotional, cognitive and sexual adaptations."[13] With this intention, the society necessarily runs up against the limits of the distinction between facts and values.

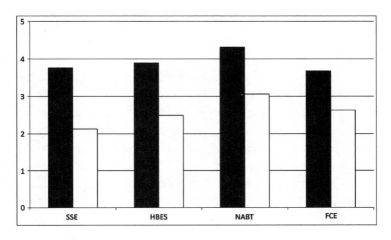

Figure 7. Response to the survey statement about a struggle for survival. Mean response to Q1 (dark bars) and Q2 (light bars) by four organizations to the statement "A struggle for survival characterizes evolution." SSE = evolutionary biologists; HBES = evolutionary psychologists; NABT = biology teachers; FCE = evolutionaries. The mean values correspond to response options in the survey: 1 = strongly disagree, 2 = disagree, 3 = neutral, 4 = agree, 5 = strongly agree. The organizations differed statistically ($p < 0.001$, Kruskal-Wallis test), as did their responses to the two questions ($p < 0.001$, Wilcoxon signed-ranks test).

Members of the fourth group, biology teachers, agreed most often with this statement about a struggle for survival. In fact, over 85 percent of them did so. This is noteworthy for a few reasons. The NABT is the largest group of American biology teachers, with more than nine thousand members, and its mission is to "empower educators to provide the best possible biology and life science education for all students." Consequently, its members act as a significant medium between evolutionary biologists and society—they provide the only education about evolution that many adults will ever have. Furthermore, the NABT represents the biologically literate American population within my survey. The social psychologist Shalom Schwartz, for example, used teachers from grades four through ten to study cross-cultural values, observing, "No single occupational group can represent a culture, but grade school teachers may be the best single group: They play an explicit role in value socialization, they are presumably key carriers of culture, and they are probably close to the broad value consensus in societies rather than at the leading edge of change. Teachers are also more numerous, literate, accessible, and receptive to research than most other groups in virtually all societies." To the extent that the NABT is an adequate proxy for the general population, the results suggest that much of the American population accepts a competitive view of evolution. In this respect, my results mirror the 1993–1994 General Social Survey, where 64 percent of respondents agreed or strongly agreed, whereas only 14 percent disagreed or strongly disagreed, with the statement "Nature is really a fierce struggle for survival of the fittest."[14]

The survey not only asked respondents to evaluate the truth of the statement about a struggle for survival (Q1, results presented above), but also addressed its value-ladenness. It did so with the following pair of questions, Q2: "Do you believe [this statement] would be beneficial if applied within society? Would it be a good thing if people were to use this statement as a guide for social practices?" Q1 sought respondents' conception of empirical support for metaphorical statements, whereas Q2 investigated respondents' sensitivity to their normative valence and to their potential application in society. By comparing responses to these two sets of questions, I hoped to understand the extent to which the fact-value distinction applies in practice—or at least within the context of a survey. A particular metaphor may naturalize diverse oughts, but the point here was simply to infer whether people detected any resonance.

In contrast to the results for Q1, the majority of the evolutionaries, evolutionary biologists, and evolutionary psychologists disagreed that it would be beneficial for the statement about a struggle for survival to be applied in society, whereas biology teachers were more likely to endorse this statement than to reject it (Figure 7). Most respondents felt that a struggle for survival characterizes evolution, but that this knowledge should not be applied to human society. As predicted by the systems ecologist Pille Bunnell, "[People] submit to the argument that nature decrees it so, but are unhappy about it."[15] Evolutionary biologists appear particularly concerned about the extension of is to ought, perhaps because in the past it led to the excesses of social Darwinism, whereas members of the HBES must walk that line constantly because of their research interest in human evolution. Either way, we appear to be trapped by the facts of the matter, even though they may well be a conceptual prison house of our own making.

An alternative explanation for these results is that people recognized that the facts of the matter should not apply to society; that is, that ought cannot be derived from is. There are two reasons this is unlikely to be the case. First, there was a strong correlation between responses to Q1 and Q2, over an array of statements about competitiveness in the survey, which indicates that respondents tended to evaluate empirical and social aspects in concert with one another. Individuals who perceived social benefits from competitive ideals tended to see evolutionary processes as more competitive; conversely, those who considered these statements about competition more factual also tended to

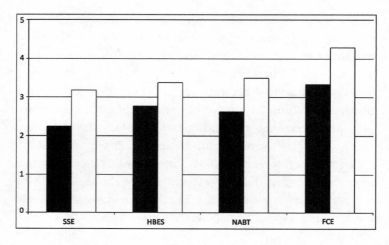

Figure 8. Response to the survey statement about cooperation.
Mean response to Q1 (dark bars) and Q2 (light bars) by four
organizations to the statement "Cooperation typifies the interaction
between animals." See the caption to Figure 7 for further details.

value them more. This result held not just for nonscientists but
for scientists too.[16] The respondents were not applying a distinction between facts and values to interpret these metaphors, but
instead treating them as fact-value hybrids. We can deny that
such transgressions occur, yet it appears that they do.

There is a second and more compelling reason for concluding that respondents were disregarding the distinction
between facts and values. If the lower positive response to
Q2 indicates that respondents objected to being asked such
questions, we would expect a similar, less positive response
regardless of the statement under consideration. But survey
respondents evaluated the statement "Cooperation typifies the
interaction between animals" quite differently (Figure 8). In-

dividuals from three of the four organizations disagreed (on average) that biological research provides empirical support for this statement, whereas members of each of the four organizations more often agreed than disagreed that it would, if applied, have social benefits.[17] Especially note that respondents evaluated Q2 quite positively here compared to the statement about struggle, indicating that they were not rejecting Q2 simply because they felt that the facts of the matter should not be applied in society. Instead, I interpret this response to mean that they were responding to the positive value resonance of cooperation relative to struggle for survival.

Respondents generally characterized evolution as a struggle for survival, so it is perhaps unsurprising that they considered animals uncooperative, even though they felt it would be beneficial if they were. The switch documented for this cooperative statement compared to the one about struggle provides evidence for a marked bias toward perceiving competitiveness in the natural world. One evolutionary biologist exemplified this bias in a written response after completing the survey: "I would have to say overall that evolution is more typified by conflict than cooperation."

Responses to two other statements reveal more fully the competitive view held by evolutionary science. Over 87 percent of evolutionary biologists and psychologists agreed that there is empirical support for the claim "There is an arms-race between predator and prey as they evolve in response to one another"; more than 79 percent of them agreed that "sperm compete with one another to fertilize an egg." In contrast, biology teachers were slightly less likely to agree with these statements, and only about 60 percent of evolutionaries did.[18] This pattern probably indicates that the concepts of arms race and sperm competition were more widely accepted among evolutionary biologists and psychologists because of their technical knowledge. They have come to think of predator and prey interactions in accord with a Cold War analogy, so that predators and prey now really *do* evolve reciprocally in response to one another in this fashion. Similarly, sperm *do* compete with one another from the perspective of these biologists and psychologists, so that whether this is an adequate way to conceive of how they interact is no longer a question.

In contrast, the evolutionaries, in particular, appeared quite sensitive to these statements, which indicates their social resonance for this group. Some scholars have argued that even the idea that sperm compete is not ideologically neutral, for it relies on cultural stereotypes of how males and females are supposed to act. Similarly, *arms race* may have had a decidedly negative valence for the evolutionaries because of its military associations, even though they were probably unaware of its technical meaning. The philosopher of biology Elliott Sober's statement that "the mother-child relationship is the setting for an arms race, in which each side evolves strategies and counter-strategies in response to the other," would be unsettling to the evolutionaries, along with many nonbiologists, regardless of whether the phrase has a purportedly technical meaning in evolutionary biology and psychology.[19]

Changing Views of Competition

Together these results provide strong evidence that people disproportionately project the concept of competition onto the natural world. Keller claimed that this is a common bias within evolutionary biology because "two tacit and complementary equations were collectively established: on the one hand, between conflict, competition, individualism, and scientific realism—that is, what life is really like—and, on the other hand, between cooperation, harmony, group selection, 'benefit of the species,' and a childish, romantic, and definitely unscientific desire for comfort and peace, benevolence and security—indeed, motherliness itself." Yet this association between competition and realism does not have to hold. For instance, evolutionary biology has harbored a tradition of examining cooperative interactions, since Darwin himself in *The Descent of Man*. Darwin explained cooperation as the outcome of selection operating at the level of groups, which Pyotr Kropotkin, in the Russian context, developed into a theory of mutual aid, one that reflected the necessity of cooperation in a difficult environment. His theory, however, had little effect on the face of evolutionary theory until recently. As summarized by the evolutionary biologist John Maynard Smith, cooperation was "largely ignored" in evolutionary biology "until the 1960s."[20]

In his book on competition, the ecologist Paul Keddy hypothesized that the focus on competition results from a combination of factors, including a taxonomic bias and, more important, the dominance of male researchers. These men not only are drawn to the excitement of competitive as opposed to cooperative interactions, but also operate within a cultural context that nurtures a tendency to compete among themselves to have successful careers. In other words, they project their individual and cultural contexts onto their studies of the natural world. Keddy wrote, "Scientists consciously and subconsciously draw upon their culture for models," insightfully observing that they "can only draw models from the possibilities of which they are aware, and perhaps ecology has been hampered by restricted access to individuals (and ideas) offering co-operative models for society and nature."[21] In a competitive cultural milieu, the challenges of considering such alternatives cannot be overstated. As this factual basis for how we see the natural world—and thus ourselves—slowly shifts, however, so might our policies and our worldview.

A number of scholars have critiqued our competitive bias as a metaphorical imposition of a particular culture. As Lakoff and Johnson explained, "Darwinian adaptation was misleadingly metaphorized by others in terms of 'competition,' a competitive struggle for scarce resources in which only the strong and cunning emerge victorious, garnering the goods necessary for life and happiness." In ecological science, for example, the idea of competition became so broad that it included cases where no conflict even occurred. Although animals sometimes directly limit others' ability to access a resource, they often simply lead their lives, but their passive use of limited resources may decrease those available for others (which is termed scramble, or exploitation, competition). To test definitively for this form of competition, ecologists exclude one species and observe whether a remaining one does better. This test provides no direct observation of competition, however—and this is also the case for the more general term *struggle for existence*. Furthermore, ecologists linked veritable scarcity with inevitable competition by assuming that resource consumption is a zero-sum game. This theoretical perspective discouraged any *actual* investigation of how organisms interact.[22]

Ecologists sought to redress the emphasis on competition by initiating a noteworthy revival of mutualism studies in the 1970s. On the surface, they decried competition as a panchreston, that is, a term that had become too broad to be usefully delimited, but this merely demonstrates its metaphorical scope, the size of the web in which it was embedded. There was nonetheless a shift toward studies of cooperation, and in reviewing this shift, the ecologist Douglas Boucher asked whether "mutualism [is] destined to be part of a *new* new synthesis, in which Newtonian ecology is replaced by a more organicist, integrated, value-laden view of the natural world." He then proposed "a programme to replace Newtonian ecology's 'competition is the basic organizing principle of nature' with 'mutualism is the basic organizing principle of nature.' Instead of being red in tooth and claw, nature is seen as green in root and flower." We see such a program, in stark contrast to the words of the evolutionary biologist survey respondent quoted above, in those of the popular systems theorist Fritjof Capra: "Life is much less a competitive struggle for survival than a triumph of cooperation and creativity. Indeed, since the creation of the first nucleated cells, evolution has proceeded through ever more intricate arrangements of cooperation and coevolution."[23]

Ultimately, however, it is misguided to ask whether cooperation or competition predominates in nature. Sober declared, for example, "Biologists now universally acknowledge that altruism can evolve and actually has done so. The picture of nature as thoroughly red in tooth and claw is one-sided. It is no more adequate than the rosy picture that everything is sweetness and light. Kindness *and* cruelty both have their place in nature, and evolutionary biology helps explain why."[24] We cannot empirically weigh their prevalence in nature in part because they are value-laden metaphors from the human realm. But we can weigh humans' preference for them; the survey provides evidence for a competitive proclivity.

We might wish to provide a counterbalance to this tendency toward competitive perspectives. For example, Susan Oyama, a psychologist and developmental systems theorist, has stated, "I am not advocating that we abandon existing methods of studying competition and cooperation, but rather that we take very seriously their limitations, and then ask, Is there anything else, perhaps quite different, that we might want to know, as scientists or as citizens?" Similarly, the biologist Lynn Margulis, founder of the endosymbiotic theory that the cells of eukaryotic organisms derive from symbiosis among prokaryotic ones, critiqued the standard way we define symbiosis as "mutually helpful relationships or animal benefits, implying social contract or cost-benefit analysis by the partners." She reckoned that "this definition is silly" because "symbiosis is a widespread biological phenomenon that preceded by eons the human world and the invention of money." As a replacement, she has proposed that we define symbiosis as "protracted physical associations among organisms of different species, without respect to outcome," which emphasizes their phenomenological interaction rather than imposing our dualistic categories of cost and benefit.[25]

Both Margulis and Oyama recognize that our adoption of a competitive worldview has social and political consequences. By balancing corporate liberalism with a more cooperative worldview, we may set ourselves more firmly on the sustainability path. In this respect, biologists encounter a significant dilemma. They tend to view biology in a competitive light, yet, like others, they perceive social benefits to applying cooperative notions that they cannot reconcile with biology itself. I am not implying that they can or should skew reality to be more cooperative; rather, I hope to have established that they have already done so, with a bias toward competitiveness that reveals preexisting metaphysical and cultural commitments rather than anything scientific. When we adopt metaphors of competition, we simply highlight this aspect of relations between ourselves and others and between ourselves and the world.

To overcome this competitive bias, some scholars even propose new metaphors for evolutionary science. We might view evolution not just as a physical fight for survival, but as a process of fecundity rather than elimination, creation rather than death and destruction. Ricardo Rozzi, a Chilean ecologist and philosopher at the University of North Texas and the Universidad de Magallanes, and his colleagues, for example, demonstrated how the metaphor of natural selection motivates competitive policies and hence justifies a continued emphasis on technological progress within Western cultures.[26] Given the association between progressive evolutionism and environmental destruction, they proposed using the revitalized, nonprogressive metaphor of *natural drift* as an alternative. In so doing they sought to denaturalize the competition metaphor. It is not just a scientific fact, but one irrevocably interwoven with social practices.

People can undoubtedly be competitive, yet nonetheless some cultures detect a fundamental generosity and kindness in human nature. The Dalai Lama, for example, has written about Buddhists' cooperative interpretation of human nature to "challenge the influential belief that somehow, biologically, we are destined to be selfish, aggressive, and violent."[27] Our competitive worldview has become largely unconscious and thus all the more powerful. Maybe we have even lived this metaphor long enough that we have truly become it; it has become a self-fulfilled prophecy. The power of competitive metaphors continues, which raises all the more questions about whether we need a new kind of science that is more cautious with the metaphors it promotes in the first place.

Value-Laden Metaphors and the Future of Science

This discussion of competition reveals how metaphors entangle facts and values. By using them scientists are opening up interpretation of their findings to colloquial meanings. Yet one survey respondent, an evolutionary biologist, wrote, "Be aware that competition means just what it means technically. . . . We use it in ecology without the emotional connotations it has among laymen." Unfortunately, we must expect these connotations, not least because most people cannot reasonably be expected to understand the precise technical meaning—if there even is one. History shows that the metaphoric flow between science and society has never stopped.

As we've seen, the value-ladenness of metaphors provides another reason to discredit the idealized notion of a value-free science. Nonetheless, this ideal persists, in large part because the fact-value dichotomy on which it depends has a taken-for-granted quality that makes it difficult to question. It implies that scientific knowledge should deal with facts alone, that is, knowledge of what the world is like without the contamination of human knowing. The historian of science Robert Proctor elucidated the origin of this way of seeing in terms of the Enlightenment split between primary qualities, which inhere in nature itself, and secondary qualities, which derive from the human subject. The latter include all our sensual experiences (sight, taste, smell, touch, hearing) and, most crucial in the current context, moral sensibility, and were distinguished from the primary qualities by their fickleness and subjectivity. Because the distinction between these qualities has been so ingrained in our culture, the fact-value dichotomy to some extent escapes critical inquiry. Consequently, Proctor concluded, "Value, divorced from the world, is also forced from the science that studies that world."[28] The invitation now is for an environmental science that brings value back in so that we might conserve our home planet.

Returning to the web metaphor, we might imagine a web

of objective concepts that are composed purely of primary qualities, separate from human knowing. Our language, however, belies that image. The language we use resonates with varied values, and it is these values that cause people to "stick" to one part of the web versus another; or, put another way, these values prevent people from relating to the web merely as a conceptual one from which they are disconnected. Though some scientists might idealize such a conceptual web, it will be made continually sticky by its interaction with people of diverse perspectives and experiences who rarely want to relate to it as merely a descriptive phenomenon as opposed to something that matters to them.

Science is imbued with metaphoric values that link it with its broader social context. This is not supposed to happen, because of a distinction between constitutive and contextual values that mirrors that between facts and values. Contextual values include social and religious influences that surround the scientific process (for instance, affecting what scientists choose to study), whereas constitutive values hold within scientific practice itself. Although there are many reasons to question this distinction, this is particularly the case when "the boundary between ethics and epistemology becomes less distinct." In particular, the Stanford philosopher Helen Longino showed that "contextual values can be expressed in or motivate the acceptance of global, framework-like assumptions that determine the character of research in an entire field."[29] Such global assumptions represent one way that contextual values may operate *within* scientific research, and they have substantial effect because the questions scientists ask, the practices they adopt, and the data they collect all bear their imprint.

For example, scientists rely on epistemic values, such as coherence, consistency, fertility, predictive capacity, and simplicity, which encapsulate the goals of scientific research, but which are nonetheless values. In the memorable words of Midgley, science elsewhere would not necessarily have adopted "the Augustan ideology that shaped the peculiar British version of the Enlightenment—that exact mix of rationalism, empiricism, Whiggish politics, Anglican theology, pragmatism and misogyny that the champions of science in that age devised." Thus, when scientists make a pledge to certain epistemic values, they may simply endorse prevalent contextual values. Epistemic values, therefore, may be considered metavalues about what constitutes good science. As the Harvard philosopher Hilary Putnam averred, "If these epistemic values do enable us to correctly describe the world (or to describe it *more* correctly than any alternative set of epistemic values would lead us to do), that is something we *see through the lenses of those very values*."[30] We have simply decided which values we allow to influence our facts.

The distinction between facts and values contributes to a specious distinction between science and society that may have ideological implications. Proctor explained: "In the mid-twentieth century, Hume's call for a separation between 'ought and is' became a rallying cry for scientists and philosophers defending the neutrality of science. Hume was used to widen the rift between the empirical world and moral obligation, one that could be used in a liberating sense to criticize church and state domination but also . . . in a more conservative sense, in defense of the immunity of science from moral and political critique."[31] Such a rift may be insidious, especially in evolutionary thought, given the extensive interplay between social and scientific conceptions of human nature. For example, a scientist may use a technical metaphor whose popular connotations affect policy, though this linkage can be easily denied. By enforcing the fact-value distinction, scientists make a political statement about the relation between science and society and simultaneously avoid more careful deliberation about the consequences of their choices. They thus avoid appropriate responsibility for choosing value-laden metaphors.

In summary, feedback metaphors allow unexamined cultural assumptions and contextual values to influence the constitutive values of scientific research practice. If these metaphors are then interpreted as facts, their values may remain hidden behind a veil and thus beyond critique. As they begin to constitute our worldview, such metaphors become more and more difficult to evaluate critically or even bring to consciousness. They may become myths that deeply influence our interpretation of the world. There is a close relationship between metaphor and myth; as Bernard Lonergan, a Canadian Jesuit philosopher, observed, "Metaphor is revised and contracted myth and . . . myth is anticipated and expanded metaphor."[32] As these metaphors become solidified through continual usage, the connections they forge between the way the world is and the ways humans should operate in that world become less open to scrutiny.

A particular metaphor may hide the possibility that we may obtain different insights into nature with a variety of metaphors rather than relying on only one. Once a certain metaphor has become naturalized, however, we may forget that it is a metaphor. Our worldview has been constituted for us, and we live according to it, whether in scientific practice or in routine activities. There is thus an important sense in which what we envision as possibility, what should be, becomes what is, or at least influences what is.[33] In this way, the metaphors that scientists choose may prophesy what is to come.

This chapter demonstrates such a process in the case of

metaphors of biological competition. They certainly have a normative dimension, one that survey respondents were quite able to detect and evaluate. This provides a cautionary tale of the consequences of our metaphoric choices. Metaphors can become self-reinforcing prophecies, in that our view of ourselves contributes to a view of the natural world that feeds back to strengthen that preexisting view, notwithstanding a presumed distinction between facts and values. Sustainability might require more responsible metaphors, including more careful consideration of the presence of prevailing cultural values.

IV
Engaging the Metaphoric Web

*Even the best-intentioned reformer who uses an
impoverished and debased language to recommend renewal,
by his adoption of the insidious mode of categorization and
the bad philosophy it conceals, strengthens the very power
of the established order he is trying to break.*
—*Max Horkheimer and Theodor Adorno,*
Dialectic of Enlightenment

*We are all connected. Metaphor knows this and therefore
is religious. There is no separation between ants and elephants.
All boundaries disappear, as though we were looking
through rain or squinting our eyes at city lights.*
—*Natalie Goldberg,* Writing Down the Bones

I n previous chapters, we have discovered how the metaphoric web entangles science and society. Recognizing
their entanglement, and not just with metaphors, numerous scholars have contended that morality ought to be
central to scientific investigation. The historians Robert Young
and Anne Harrington proposed a "moral science" and a "compassionate science," respectively. Feyerabend endorsed ethics
as an "overt judge" of scientific truth, and another philosopher,
Columbia University's Philip Kitcher, proposed a "responsible
biology" in which ethical considerations are integral. Summarizing the need for a new form of science, the Canadian
historian Stephen Bocking recommended that "a more democratic science is necessary in order to question directly the often
unexamined imperatives embedded within science: to reshape
the world, to impose a standardized perspective, to pursue efficiency above all else."[1] These proposals are part of a more general trend toward citizen engagement in postnormal science, as
mentioned earlier, which recognizes that science is not necessarily progressive and thus must be more accountable to society. Here I wish to examine how metaphoric choices might fit
within such proposals for a postnormal and responsible science.
How can environmental metaphors be true to both science and
society, to both facts and values? I provide a broad theoretical
perspective on such questions in this chapter, before examining
two more case studies; then, in chapter 7, I address more specific
questions about how responsible metaphorics, including citizen
involvement, might operate in practice.

Numerous scholars have underscored the need for greater
reflection on the social implications of scientific metaphors, as
we saw in chapter 1, yet they may remain unaware that their
efforts fall within the mandate of the field of critical discourse
analysis. This field investigates the instantiation of power and

hence inequality through language and commits itself to bringing about social change. It recognizes the dialectic between objects in the world and concepts in our discourse: the language we use is shaped by the society in which we live at the same time that that society is shaped by the language used. It follows that those who control language have the potential to covertly dominate society by naturalizing particular ideas. Critical discourse analysis encompasses the subdiscipline of critical metaphor analysis, which focuses on the relation between metaphor and ideology in spoken and written texts; environmental and political themes are prevalent. The apposite field of critical ecolinguistics provides critical insights too by taking a broader perspective on the relation between language *systems*—rather than just words—and environmental issues.[2]

These critical studies emphasize that metaphors can be assessed in various ways, not only for their epistemic merits, but also, for example, for their social and environmental ones. Harré and his coauthors distinguished three ways to assess the adequacy of language with respect to environmental affairs. The first is *referential adequacy,* which questions whether a language has the "lexical resources to discuss a topic in sufficient detail, 'sufficiency' being relative to the task in hand." This concerns whether linguistic resources are misleading, vague, or even nonexistent. Among other examples, they discussed the vagueness of the phrase *greenhouse effect* and Rachel Carson's reframing of insecticides as biocides. They noted that advertisers add the prefix "eco-" to help sell their products, and they observed that some languages have a prefix that indicates harmfulness and permanence, one that we lack, along with words for "not biodegradable," "a positive weed," and others. More generally, English words promote commercial exploitation of the environment when they appear neutral but their connotations are

harmful (for example, *develop* and *resource*); when they give a
pleasant name to something unpleasant (in euphemisms such
as *improving nature, clearing, harvesting,* and even *global warming*); or when they give something neutral or even positive a
pejorative slant (such as "overmature" trees).[3]

It is the two other types of adequacy that they discussed,
however, that are more pertinent to this book. A language that is
socially adequate should be "acceptable to a maximum number
of speakers in the target community, promote social unity and
intercommunication and cater to present as well as anticipated
future social needs."[4] Their example is germane. They showed
that the language used to accentuate the problem of human
overpopulation includes terminology such as *population explo-
sion* and *population bomb.* It follows that the proposed solution is *population control,* terminology that many people find
jarring because it sounds too much like birth control and pest
control. For some women, in particular, this terminology may
raise concerns about whether they will lose responsibility for
their bodies to external agents of control.

Finally, a language that is *environmentally adequate* should
also "enable its users to talk about environmental matters in
an informed manner and promote the well-being of its speakers and nonhuman nature."[5] This reflects the ultimate concern
of this book, which is whether the metaphors we use in environmental science nurture sustainability. I assume that this
category to a large extent subsumes the previous one, for if
our language is to be environmentally adequate, it must also
be socially adequate.

This proposal to evaluate the socioethical dimensions of
environmental metaphors may seem quite radical and perhaps
unfeasible—and we will examine the latter question in chapter
7. But it is not that much of an extension from what scientists

already do. I take it as a given that many environmental scientists are committed to environmental causes. In interviews with eighteen prominent ecologists, for example, sociologists found that "many . . . stated that they felt a desire to 'save the planet' or prevent environmental damage."[6] This finding echoes the call among scientists themselves for research that has greater relevance and thus influence. Or we could go further and, following Kitcher, assert that the primary objective of science should be the common good, one element of which is the environment. Either way, we have to expand the usual emphasis on epistemic dimensions of scientific terminology, where a metaphor would be abandoned only if it is unproductive or inaccurate. This is not enough in the domain of environmental concerns. We cannot just assume that the truth will set us free, not least because such truths are often couched in socially resonant language, but also because cognitive science shows that if data fail to fit people's frame, they will reject the data rather than abandon the frame. Thus, we need to contemplate social and environmental dimensions and explicitly evaluate them where possible.

Westoby modeled reflection on scientific language along social and environmental rather than just epistemic lines. Although ecologists are taught to discriminate is from ought, he reviewed research demonstrating that nonscientists nonetheless look to ecology for moral and normative guidance about environmental issues; they see it as a secular religion. In response, he first recommended education about the alternative quasi religion of research and the distinction between science and morality, but he realized this may have limited capacity to resolve the tension between what academics and the public want from ecology. Recognizing that ecology has escaped from academic control so the "'general-public' meaning . . . will inevitably prevail as the language evolves," he suggested that in some contexts ecologists should adopt different words in their texts and courses as a defensive option. He concluded, how-ever, by considering how they might instead engage rather than retreat by selecting "sound generalizations from which moral and aesthetic principles could be permitted to flow," providing metaphor-rich examples such as biophilia, evolutionary ma-chinery, and shifting mosaics.[7] Despite the difficulty of selecting sound generalizations from among these and other options, I think his question directs us toward a more socially respon-sible form of biology. Although scientists will never find perfect metaphors, from both epistemic and social perspectives—or the royal road to sustainability—the intention to choose carefully may itself prove formative of a new relation between science and society. My intent here is not so much to suggest appropri-ate generalizations as to consider some of the themes by which we might evaluate environmental metaphors.

It is worth exploring briefly the consequences of taking Westoby's ideas to a logical endpoint. People will continue to take broader meanings from science, in part because of feed-back metaphors. Thus, like it or not, science interfaces with re-ligion, though some would go much further and maintain that, as a meaning-making exercise for understanding the world, science is fundamentally religious. The Latin root of religion, *religio,* means "to bind," and scientific insights can help us be-come bound to and to care for the world. If this is the case, we must explore this interface between religion and science. As the eminent English philosopher Alfred North Whitehead stated in 1925 in *Science and the Modern World,* "When we consider what religion is for mankind, and what science is, it is no exaggera-tion to say that the future course of history depends upon the decision of this generation as to the relations between them."[8] Metaphors help to form this relationship, so Westoby's reflections provide a timely precedent for studies of their role and their significance.

I contend that the common good that we seek with our environmental metaphors is sustainability. This is what these metaphors are now called to do, to make the world meaningful to us. "To sustain" is to enable something to endure. Applied to the environment, the idea of sustainability involves our seeking a future in which the basic needs of humans are met, but without impairing (or destroying) the natural systems and species that support us. Nonetheless, sustainability is itself a difficult and contested concept, which raises questions about its referential adequacy. Without reviewing its history or the extensive debate over its definition, I side with recent authors who conclude that we can agree on many of its basic tenets and that even its ambiguity can be a source of creative tension between stakeholders.[9] It still retains importance, though we must ensure that the concept ultimately accomplishes something.

We might aim to sustain in various ways. Metaphorically, sustainability may be the journey or the destination, and there are both many ways to journey and many possible destinations. Often, we hear of the model of a three-legged stool of sustainability, with its biological, economic, and social legs (the three P's: people, planet, and profit). A major limitation of this model, however, is that it aligns with traditional disciplines and institutions, which may inhibit our ability to derive truly innovative solutions. It implies that we have separate legs rather than fundamentally interconnected elements. There is also a very real risk that the economic leg will predominate so that suddenly the stool is off-kilter. Perhaps we need to be doing much more, elevating the severity of the issue with metaphors that query whether our current actions and patterns are even survivable.

We can, I hope, agree on some of the core principles of

sustainability. These include its challenge to conventional thinking and its emphasis on context, equity, and interdisciplinarity.
Each of these underlies my approach to linguistic socioecological sustainability, though some of this will not become clear
until chapter 7. I seek language that challenges the status quo, in
particular the idea that we just need to be more efficient to attain sustainability rather than making more radical adjustments
in our thinking about the systems in which we are embedded.[10]
Methods of public participation must occur in local contexts,
and they have much to do with promoting equity, by including
not only diverse human perspectives, but also other species.
And finally, I seek to break down the barriers between ecology
and society that can prevent sustainability, which in particular
means that we require insights from various disciplines. I also
adopt sustainability here for the usual reasons that it is adopted:
its flexibility and the extent to which many of us recognize and
relate to it. Rather than certainty, it provides a vision of possibility. In that sense, sustainability progress is a journey that we
embark on together rather than a specific destination.

Language and Worldview

Before continuing, I wish to draw attention to the underac-
knowledged relevance of language in seeking sustainability. Although I touched on this in chapter 1, it is quite pertinent to my
overall argument, so I wish to deepen that discussion here. In
seeking a more enduring relationship between humans and the
ecological systems on which our well-being depends, we need to
consider not only the direct relationship between humans and
nature, but also the intermediary of linguistic representation.
Humans operate to a large extent through linguistic symbols,
which interpret the world in particular ways that have associ-

ated performative consequences. Metaphors, as I have shown repeatedly, provide a quintessential example through the large role they play in constituting our worldview.

To engage the metaphoric web, we have to appreciate the power of language as a means toward sustainability ends. We live in a sea of language, and thus we may forget its influence on how we traverse the waters of life. It is challenging for us to examine this linguistic link between ourselves and the natural world, for it is like fish reflecting on the water in which they have lived their entire lives. We cannot escape language to look at it. In response to this quandary, the quantum physicist David Bohm created an imaginary language, which he called the rheomode, to highlight the fragmentary Newtonian worldview created by ordinary language and the possibility of an alternative that more adequately reflects process-based modern physics.[11] Some critics would contend that this gives too much weight to a particular interpretation of quantum physics, perhaps proposing as an alternative that we experiment with less language altogether, as some mystical and contemplative traditions have done.

Instead, to understand the effect of our language, we might turn to the evidence provided by the thousands of other languages spoken around the world. We know from everyday experience with speakers of other languages that these languages differ from one another, but to what extent does this actually influence people's conceptual systems? Some linguists claim that language has a dramatic effect. In the noteworthy words of the American linguist Benjamin Lee Whorf, "We dissect nature along lines laid down by our native languages. The categories and types that we isolate from the world of phenomena we do not find there because they stare every observer in the face; on the contrary, the world is presented in a kaleidoscopic flux of impressions which has to be organized by our minds—and this means largely by the linguistic system in our minds." In its strong form, linguistic determinism, this effectively means that language determines reality. Through extensive argumentation and experimentation, this thesis has been largely discredited, though a weaker form that contends that language *influences* our perception, the thesis of linguistic relativity, has withstood at least some empirical testing.[12]

A recent review by the Stanford psychologist Lera Boroditsky, for example, demonstrates how language affects cognition about space, time, objects, and substances, thus suggesting that "the private mental lives of people who speak different languages may differ much more than previously thought." Consider the case of grammatical gender, those languages such as French, German, and Spanish (unlike English), in which nouns are assigned a gender. Boroditsky conducted experiments showing that the grammatical gender given to an inanimate object causes people to think of it as having that gender. The results were as follows: "Asked to describe a 'key' (a word masculine in German and feminine in Spanish), German speakers were more likely to use words like 'hard, heavy, jagged, metal, serrated, and useful,' while Spanish speakers were more likely to say 'golden, intricate, little, lovely, shiny, and tiny.' To describe a 'bridge,' on the other hand (a word feminine in German and masculine in Spanish), German speakers said 'beautiful, elegant, fragile, peaceful, pretty, and slender,' while Spanish speakers said 'big, dangerous, long, strong, sturdy, and towering.'" As another example, Russian speakers discriminate between dark and light blue with different words, and experiments show that they thus discriminate more rapidly between shades across these two color categories than within them (whereas there is no difference for English speakers).[13] These and other examples reveal the intricate connections between language and thought.

Consider some of the partial ways that English structures the world. This exercise will exhibit features that might otherwise remain unexamined (except among linguists), thereby reminding us that the world is not a given, but that we perceive it in a particular way that is, to a variable extent, influenced by the language we speak. The sociologist Saroj Chawla, for example, asserted that the "linguistic and philosophical roots of our environmental crisis" lie in three features of the English language. First, English individualizes what are known as "mass nouns." Individual nouns refer to items with a distinct outline, such as a car, a human, or a rock, whereas mass nouns "denote a homogeneous continua without implied boundaries," such as water or coffee or beer. In English, we individualize mass nouns when we refer to a glass of water, a cup of coffee, or a mug of beer. We also delimit mass nouns within measurable categories, such as referring to the volume of a liquid in liters. Though such measurement has benefits, it simultaneously fragments rather than emphasizing the more holistic sense that all water is connected. Chawla proposed an implication of this: "As long as we think of the water in the home and the industrial waste water in the rivers or ocean as distinctly separate, it will be difficult to avoid water pollution."[14] On a related note, we stress biological classifications and scientific distinctions rather than the wholeness of life. If you imagine the difference between sitting in a forest, hearing the leaves high overhead rustling in the breeze, and thinking of the same forest in terms of board feet, you get an even better feel for the distinction.

Second, English allows us to count two types of nouns, both real and imaginary ones, even though the former are "perceptible spatial aggregates" (such as cars, humans, rocks), whereas the latter are "metaphorical aggregates" (such as happiness and well-being). Not all languages quantify the latter, thereby attempting to "objectify and measure experience." To the extent we do so, we may forget that what is meaningful may not be countable and that we may not need more of what is countable (money, possessions, and so on) to obtain those things that really matter.

Third, and finally, English relies on a fragmented three-tense conception of time. In speaking of past, present, and future, distinctions that other languages may ignore, we objectify time and it becomes quite linear, each unit (seconds or days or years) equal to every other and stretching both backward and forward into the distance. Chawla explained the result: "Years, centuries, and decades are nouns; they are pluralized and enumerated as if they are touch-and-see objects. In this way, subjective experience of real time has been lost."[15] One result of this conception is that we are hoodwinked by the ideology of progress, one that can exist only when we imaginatively place distant past and future time in front of us as if they exist in the same way as the present moment.

Chawla and others have concluded that English is ill-fitted for a sustainability ethic because it is quite fragmentary, emphasizing things rather than processes. There is no need to invent a new language, a rheomode, however, because other languages provide pointers. As a specific example, linguists have argued that the grammar of Niitsi'powahsin, the language of the Blackfoot peoples, is more consonant with the process-based worldview of modern physics cited above, not to mention its consonance with the worldview of many process philosophers and even major world religions such as Buddhism. To summarize: verbs dominate in Niitsi'powahsin, so "nouns seem to be verbalised out of existence." Nature therefore becomes "a flux of processes."[16] This foregrounds relations over objects, providing a more ecological understanding that may be better suited to giving precedence to our relation with our home planet—and to transforming this relation. The question is not so much whether this view is closer to the truth, for it is simply another way of conceptualizing experience, but whether we might adopt from an alternative model such as this one if it is more apt.

These language differences become all the more significant when one recognizes the extent to which English is the language of contemporary science, including environmental science.[17] It follows that the limitations we have just seen greatly influence how we conceive of environmental problems; we see them largely as problems out there in the objects studied by science. The solution thus lies in science itself because it provides a better understanding of the facts of the matter. The worldview associated with English thereby reinforces the attitude of scientific realism, which tends to neglect all the other ways we could conceive of and relate to our experience. Again, the emphasis is on things, understood by reduction, rather than processes. We might agree to explore the world from diverse cultural starting points, but science too often becomes scientism, the belief that it is the one valid method of inquiry. Instead, different sources of cognitive diversity allow cultures to inhabit landscapes and settings in variable ways, and this leads to different metaphors that shape what is of interest to them.[18] The Western scientific approach has both benefits and shortcomings, and in the interest of sustainability we need to draw on every source we can rather than relying so exclusively on science, a source that, in particular, has denounced the incorporation of values.

We might learn from those relations between humans and landscapes that have evolved over eons in situ and that do not require facilitation from natural science. For example, in his beautiful and brilliant ethnography of the Western Apache, *Wisdom Sits in Places,* Keith Basso, an anthropologist at the

University of New Mexico, described just such a place-based relationship. The local landscape of these peoples is densely packed with specific places whose place names (toponyms) invoke rich historical narratives. These narratives are not simply historical, "for as persons imagine themselves standing in front of a named site, they may imagine that they are standing in their 'ancestor's tracks.'" Thus, when an elder "speaks with names," it is "interpreted as a recommendation to recall ancestral stories and apply them directly to matters of pressing personal concern." These stories carry moral meanings that are "a call to persons burdened by worry and despair to take remedial action on behalf of themselves."[19] At other times, when people act immorally, a storyteller "stalks" them with a tale told to a larger audience, though the perpetrators know it is intended as a lesson for them. It makes them think and inspires them to transform themselves or to undo wrongdoings. In this way, the landscape of the Western Apache is imbued with cultural meaning, a linguistic link that is essential to their culture.

Disconcertingly, linguists have disclosed a rapid loss of languages around the globe. As David Harrison of Swarthmore College stated in the opening line of *When Languages Die,* "The last speakers of probably half of the world's languages are alive today." There are only a few thousand speakers of Niitsi'powahsin, for example. Over 40 percent of the world's languages are endangered, a greater rate than among the planet's plant and animal species, the focus of conservation biology. But with language loss we are arguably losing much more, including "long-cultivated knowledge that has guided human-environment interaction for millennia. We stand to lose the accumulated wisdom and observations of generations of people about the natural world, plants, animals, weather, soil, and so on. The loss will be incalculable, the knowledge mostly unre-

Figure 9. The Kallawaya of Bolivia possess an exhaustive knowledge of Andean plants and their curative uses. Illarion Ramos Condori (pictured), an expert herbalist, is among fewer than one hundred people who still know the rapidly disappearing Kallawaya language and its unique plant taxonomy. Photo courtesy of K. David Harrison.

coverable."[20] Much of this knowledge exists only in memory because many of these languages have not been written down. They contain intimate knowledge of particular places and the organisms living therein; as they disappear, we lose alternative ways of knowing, not just knowledge for pharmaceutical purposes (a common focus of bio-prospecting), but ways of being on the planet, entire phenomenological life worlds that might promote more sustainable ways of living (Figure 9). Specifically referring to metaphors, Peter Mühlhäusler, a University of Adelaide ecolinguist, bluntly stated, "As the boundary between literal and metaphorical is again language-specific and as access to reality in the sciences is achieved mainly by means of metaphor, a greater knowledge of non-Western metaphor systems could prove a significant asset." That might be an understatement, for with the loss of these languages, we lose alternative hypotheses for relating to our world. In conserving the natural world, we simultaneously maintain a rich source of metaphors, both poetic and scientific—reasoning that might seem circular, self-serving, and anthropocentric, but which instead manifests how our lives are much richer, both materially and symbolically, when we maintain as much as we can.

Evaluating Environmental Metaphors

Having reviewed the importance of language for sustainability, let us now consider some specific axes for evaluating environmental metaphors. As we do so, I hope that the four feedback metaphors I have chosen for this book will appear less ad hoc; they represent some of the great forces influencing our lives and ecological systems at present. Because of the complexity of the Earth as a large-scale system, we require one metaphor or another to conceptualize it. It is worth reemphasizing, in this context, that the metaphors we choose trade off differing benefits and costs from the perspective of sustainability. If we conceive of nature as a resource, for example, then we are likely to act toward it in that way, that is, assuming it is there for our use. In contrast, if we relate to nature as home, we may be more respectful and think more concretely about how we inhabit places. We earlier contrasted metaphors of mechanism versus personification. Clearly, differing metaphors provide significant models for how to inhabit and act as earthly inhabitants. And I suggest later that we require not one metaphor but a suite of them to fully appreciate the complexities involved.[21]

In evaluating metaphors, I maintain that a critical question is whether they simply bolster the status quo or help us begin to question it. Scholars interested in the social efficacy of language commonly evaluate whether it empowers people and contributes to their freedom, what might be called "emancipatory discourse." For example, Otto Santa Ana, a scholar of Chicana and Chicano peoples at UCLA, seeks "insurgent metaphors" to contribute to their liberation in the United States. Environmentalists and environmental scientists concerned with the state of our planet face a challenge similar to his, so they would be wise to heed his conclusion: "If history is a succession of metaphors, then they are the principal instruments by which vocabularies are created to speak society into existence. Insurgent metaphors are tools to construct stronger vocabularies to speak this new society. To contest the current regime of discourse requires the creation of insubordinate metaphors to produce more inclusive American values, and more just practices for a new society." The challenge may be understood by thinking of existing metaphors as dominant memes with which we have been indoctrinated. Our challenge, then, is to "wake from this meme dream."[22]

If environmental science is to contribute to a fundamental rethinking, we have to question many of the metaphors we adopt that represent the age. They are adopted in part because they are culturally prevalent and they sell, but in a sense they represent the lowest common denominator. Some critics thus declare that we need to shift our environmental metaphors. Among others, Richard Underwood, in 1971, called for a poetics of ecology that would provide more radical metaphors. He felt that people had lost touch with the being of life, and that our metaphoric choices have contributed to the antagonism of man versus nature. He thus viewed the environmental crisis as one primarily of metaphor, its solutions to be found in new metaphors. A few years later, the feminist literary critic Annette Kolodny advocated reforming our language to affect our relation to the land. Though not focused on scientific metaphors specifically, her work is part of a broader feminist current that traces how the feminization of landscapes, as a virgin to be admired or conquered, a mother, or even a mistress, authorized the American colonists' destruction of it.[23]

Such proposals for environmental metaphors cohere with aspects of green political thought that seek to base politics on something other than profit. Such thought endeavors to problematize the implicit bias toward industrialism in modern Western culture, especially if this habit is "ecologically irrational." In this context, one of the greatest political breakthroughs of environmentalism has been to sanction a new "forum for communication, a green public sphere. Even with its many internal differences and disagreements, the emerging green public sphere poses a challenge to the once comfortable framework of industrialist discourse." In this sense, ecology could become a "subversive science," as a result of not just the "shifting ground of particular findings, but from orienting metaphors that challenge the presuppositions of the administrative mind."[24]

Questioning the status quo means more explicitly questioning our orthodox view of the environment. In particular, scientists have identified a number of problems for which we now seek largely technological solutions. This pattern may in part derive from the very root of the term *ecology*, the Greek *oikos*, for household, which implies that we might manage this home. This managerial framing, however, is reliant on science, and it often remains unquestioned. Yet it has at least three limitations.[25] First, it trivializes the public's role, when its members are often the first to recognize environmental problems, especially in the form of nongovernmental organizations. Second, it inflates the role of science, overlooking the construction of facts and the presence of continual uncertainty. It assumes that we simply need to figure out all the pieces, scientifically, and then we can solve the problems, when in fact there are many circumstances in which science does little to resolve environmental controversies. Third, managerialism of this sort leads to standoffs between competing interests, conflicts that in some cases are entrenched by our initial framing of the situation. The upshot is that we emphasize management, a metaphor that speaks to a culture of control by the experts. To put this in context, it is worth noting that the Swedish language does not have a word for management, instead emphasizing a less control-oriented metaphor of caretaking. This raises the possibility that, as the sociologist John Maguire put it, "our modern managerial culture is itself the disease to which it claims to be the cure."[26] We might focus instead on enhancing relationships so that they are not so hierarchical, facilitating the ability of local people to solve their own problems with the assistance of science.

The dominant managerial approach to problem setting stresses a mechanistic and reductionist approach to finding solutions. As I discussed earlier, we commonly understand life as a mechanism, and biologists often seek mechanistic explanations in their research practices. Through this process, organisms may become merely machines (or perhaps parts within larger ecosystemic machines), so we act toward them as such. It is this move that we have to assess in a socioecological context. For example, researchers may feel little remorse in killing organisms because they are simply mechanisms (or automatons), like a watch, and have no emergent property called life anymore. Biodiversity scientists often kill organisms to catalogue them, but they justify their doing so by placing abstractions about biodiversity and its mechanistic function above those concerning the value of individual lives—an ethic that relates in part to viewing these species as mere mechanisms rather than as valuable entities in their own right. One of the great ironies here is that we reduce these systems to Cartesian mechanisms, yet we then state that somehow this unpurposeful stuff does matter and must be conserved.

Speaking more generally, the Georgia Tech philosopher Bryan Norton provides important direction for assessing metaphors in the context of sustainability. He promotes a philosophy of adaptive ecosystem management, and as part of this he emphasizes the need for a postpositivist ecology. To a large extent, his vision is based on critiques of the fact-value dichotomy that parallel those in chapter 3, and he specifically draws on the example of self-reflexive ecological modelers who now carefully attend to the role of values in the construction of their models. They recognize that all models rely on human values and metaphors, that we can adopt alternative metaphors from which to develop models, and that the choice among them needs to occur in the context of broader public discourse. Not only that, but we can combine this interest in metaphors with a philosophy of adaptive management by "experimenting with different metaphors and 'models' to characterize a problem" in order to choose "appropriate models for communicating about, and working to solve, environmental problems."[27] Although I fully endorse such an approach, I now turn to a particular critical axis for evaluating proposed environmental metaphors.

Connecting Nature and Culture

Although chapter 7 will consider in more detail how we might, in line with Norton's proposal, incorporate local deliberation into the selection of metaphors, I wish to focus on a general theme for new metaphors here. This theme is by no means the ultimate answer, but I present it as a counterweight to the prevailing view. Specifically, I contend that we need to focus on whether the metaphors we adopt in environmental science tend to reinforce a problematic dichotomy between nature and culture. This dichotomy reflects those between science and society and between facts and values that we have considered earlier. Science studies the solid facts of nature; society deals with the shifty values of cultural and individual preferences. In the long run, we must overcome such thinking and the metaphors that contribute to it in the interests of sustainability. As many have claimed, in various voices, we must embrace a post-Newtonian ecology that integrates humans in nature.[28] New metaphors will assist us here, encouraging fresh ways of being that are facilitated by novel self-fulfilling prophecies. At the same time, they contribute to a reformulation and critique of the long-standing fact-value and science-society dichotomies.

We might begin by contrasting our metaphors with those found in other cultures. The risk here is to romanticize indigenous peoples when they are not necessarily more environmentally benign. Nonetheless, we can learn from them, not least in greater awareness that possibilities exist other than the one we live in. The English language tends to amplify and reify the distinction between nature and culture. This distinction is hierarchical and tied to analogous and parallel dichotomies between body-mind, emotion-reason, and woman-man. In contrast, some languages lack a word for nature. Some hunter-gatherers take "the human condition to be that of a being immersed from the start, like other creatures, in an active, practical and perceptual engagement with constituents of the dwelt-in world."[29] Such a view might remind us that in taking care of nature, we are taking care of ourselves.

Nurit Bird-David, a cultural anthropologist at the University of Haifa, for example, compared the humanity-nature metaphors used by tribes around the world, including sexual, procreational, adult-child and "relatedness" ones, drawing from the Cree, Bushmen, aborigines, and pygmies.[30] She showed how the Cree, when hunting, enact a sexual relation between hunter and hunted, and how the !Kung take on the persona of a namesake species. On the basis of these and other examples, she concluded that we in Western societies inhabit a subject-object frame: nature is a resource that we need to do something with. These tribes instead relate to nature within a subject-subject frame, their emphatic emphasis being on relation. Such a relation results in part from experiencing the land firsthand rather than relating to it abstractly and as a mere conceptual problem.

Another example will help highlight how our language separates us from nature, creating varied paradoxes along the way. In the context of the James Bay II hydroelectric proposal in northern Quebec, Canada, the geographer Randy Bertolas interviewed Vermonters, nonnative Quebecers, and native Cree peoples to uncover their understanding of the concept of wilderness. He was concerned that nonnative people would expropriate the land through their culturally specific definition. He noted, "Few things define the Cree more than their ties to the land and its animals and, not surprisingly, the Cree language contains no word or concept of land that is comparable to Western notions of 'wilderness.'" Even though the concept of wilderness as an entity distinct from themselves was not known to their culture, they nonetheless revered the wilderness itself. They tended to associate it with a state of mind, at the same time acknowledging its usefulness, whereas the other groups—especially Vermonters—were more likely to associate it with an absence of humans and human disturbance. Accordingly, for these other groups *wilderness* is an abstraction representing places that are valuable for their potential resources or as escapes, whereas it was the very sustenance and context of Cree lives—it made little sense to separate themselves from it, so they described it in much more personal ways. Ironically, Bertolas found that the Vermonters "perceived no discrepancy between their humanless conceptions of wilderness and their desire to visit it (thereby impacting it)."[31] Yet such discrepancies are written into the very dualism between nature and culture that underpins so much of how we think about the environment.

Furthermore, we often frame environmental problems as ones related to nature. Hence, science becomes savior and solution, as if the problem really lies in what is "out there" in the objective world that we can approach through natural science, rather than "in here," in our selves, in our social world, and in how we relate to the world. Environmental scientists are not necessarily trained in such "subjective" issues, for they are often taught to ignore humans entirely. It is a fundamentally realist view of the world, and though undoubtedly helpful at times, the aforementioned examples from other cultures may help us see that the world can be viewed in very different ways. We cannot just assume that social and sustainability benefits will follow from particular biologically framed problems and solutions, but instead need to interrogate where they might be limited.

As one example, many environmental scientists now spend their lives in front of a computer screen analyzing data and running models, unable to identify the organisms outside their window or relate their life stories. Their focus on abstractions can contribute to a dichotomy between their concepts of the natural world and the lives of those who experience it on a daily basis. They emphasize the objective problems given priority within a small community of scientists. To the extent this is the case, such research will be limited because it underscores a tendency to think in generalities rather than the specifics of locales. And yet, many environmental scientists at the same time want to encourage a relation with a landscape from which they are in many instances quite divorced.

There is a tendency to believe that solutions must be based in Western science. Chaisson, for example, drew on cosmic evolution as a "powerful and true myth." Numerous writers have critiqued such large-scale ideologies because they emphasize our particular way of knowing—our metaphors—and close us off to those of others. Philosophers, too, fall prey to this trap. In his discussion of multicultural environmental ethics, the philosopher Baird Callicott, from the University of North Texas, considered the case of the Siberian crane and its migration through regions inhabited by people who follow about half a dozen different religions.[32] Presumably, these religions would come to different decisions about how to relate to and conserve the crane. Faced with the very general problem of how even to get some of these religions to speak to one another, let alone adopt a similar worldview, he recommended that they agree on a basic or universal environmental ethic, one that he adopted from the natural sciences. Specifically, he drew on the Epic of Evolution project as a scientific grand meta-narrative to be mediated through the diverse modes of religious representation, thereby giving the universal a local flavor. He presented little evidence, however, that it would appeal to people of these diverse religions. He acknowledged that it would need to be popularized, perhaps learning from the success of religions through history in using resonant images to capture attention. Nonetheless, his solution is our worldview—not to mention that many scientists would debate its validity. Instead, I maintain that there will have to be negotiation all the way down, on all levels, and in local spaces (such as this Siberian crane case), rather than thinking we can decide on a priori universals.

If thought about more radically, metaphor can be very helpful here. It might seem surprising, but poets themselves have debated whether their perspective is sometimes too realist. In proposing a reformative metaphorical ethics, the English scholar and poet Adam Dickinson, for example, called into question that realist commitment, showing how metaphor can point us beyond language because it is the "articulation" between presence and absence, language and nonlanguage, and ultimately is and is not. He contended, "Things cannot be captured in idiomatic realist language games," and thus metaphor is critical because it "subverts the totality of a realist perspective that argues for a proper linguistic representation of matter." In other words, literal language cannot capture our experience, despite its giving the impression that it can do so. Metaphors are important because they remind us that many of the distinctions between domains that we assume may be misleading: "interpenetration and connectedness" are at least equally valid characteristics to consider.[33] In focusing on the literal, however, we emphasize how things are distinct, and thus set off one domain from another, the literal true representation from an aberrant metaphorical one.

From Dickinson's perspective, metaphor thus becomes an epistemic tool of a different sort, one that reminds us of the fundamental interconnections between things. This extends across nearly all of the dualities we take for granted: fact-value, science-society, literal-figurative.[34] In the current context, environmental metaphors might highlight the unavoidable entanglement of nature and culture, especially in the way that they describe nature through cultural lenses and culture through natural lenses. With metaphor we see one thing in terms of another, and the key question I attend to herein is whether we are choosing the right thing—a question usually asked solely along epistemic lines, but one that can also be asked more broadly. And in seeking metaphors of sustainability, one way to assess these other things is whether they enhance our sense of interconnection.

Seeking a Language of Connection

To reformulate our priorities toward sustainability, we have to focus on how we relate to other organisms and individuals. Here the life sciences can play a crucial role, as they have lately contributed to a reconsideration of who we are. Evolutionary and molecular biology have shown that we are connected to other species historically, and ecology that we are connected environmentally. Although these new views are still metaphorical, they contain lessons that could help transform our relation to one another and to other species. Many biologists have devoted themselves to communicating these findings and to conserving ecological systems, yet some of their language fortifies a dominant discourse that is actually counterproductive to their interests. A new relation to our planetary home may require a shift in worldview, and concomitantly, in our language. What metaphors might biologists invoke to communicate a new way of relating to the world around us? Which ones would be in the best interests of society and other species? I propose that biologists need to utilize language that reinforces an understanding of how organisms are interwoven evolutionarily in the environment they create together. Metaphor provides an inroad to this understanding, so the ones we choose matter. If the world is built on relationship, then we require language of connection. I wish to draw attention to two crucial elements of interconnection. First, we want the metaphors we use to connect us to the world, in particular the other-than-human world. Second, we want the metaphors we use to connect us with one another. These are critical elements of sustainability that metaphors can provide. By emphasizing relationship, such metaphors exemplify what has been called an ethic of partnership, as opposed to former ethics based on egocentrism, anthropocentrism, or even ecocentrism. This new ethic gives equal moral consideration to both the human and the nonhuman, thus balancing respect for biodiversity and cultural diversity.[35]

Beginning with the former relationship, our connection with the world, Evernden argued that there is only one relationship that is "relevant to a discussion of man and environment," and that is "the relation of self to setting." He did not mean simple causal connection, but fundamental interrelatedness, which we will recognize only through fundamental questioning of the subject-object dichotomy on which much of how we approach the world is based. The result would be a deepseated realization that discrete entities are illusory. He provided a number of examples to defend such a claim, including symbionts, colonial organisms, and the process by which independent chloroplasts were imported into plant cells and mitochondria into animal cells. Recent research on extrachromosomal elements suggests that horizontal gene transfer between separate individuals and even species—"Creatures can 'infect' each other with evolutionary transformations"—has been a common feature of life. Similarly, consideration of the human microbiome project demonstrates that we may be more appropriately seen as a metagenome, as bodily ecosystems or superorganisms.[36] Ecology *can* be a subversive science, but only if its basic premise of interrelatedness is fully understood.

All these examples elide our skin boundary, which relates to the earlier discussion of how personification breaks down self-other boundaries. The fact-value dichotomy partly derives from assigning facts to external referents and values to internal ones, which relies on our projection of an intervening boundary between self and nonself. Although we may perceive an obvious boundary there—the skin, which undoubtedly exists—the question is whether there is any reason to highlight this boundary rather than the tremendous flux across it. Those of us inculcated in Western social values operate on the basis of this subject-object dichotomy, but, using studies of diverse peoples around the world, anthropologists have shown that such an individualistic view is in fact peculiar. As the American poet Mary Oliver put it,

> I'm never sure
> which part of this dream is me
> and which part is the rest of the world.[37]

The possibility that we are fully part of the world raises further questions about the contemporary dismissal of personification in science, especially when we do it anyway. The more critical issue may be the specific types of values we invoke with the forms of personification that we choose, given the long-term political and social implications of our metaphoric choices.

The Cambridge philosopher of science Mary Hesse once stated, "Rationality consists just in the continuous adaptation of our language to our continually expanding world, and metaphor is one of the chief means by which this is accomplished." As we become more aware of interconnection through scientific exploration in both microscopic and macroscopic domains, we might adopt new metaphors that better communicate this sensitivity. Evernden concluded that the challenge for ecologists is that "many of the most significant arguments [in the environmental movement] cannot be handled by their lexicon," but that does not have to be the case.[38] Instead, a new emphasis should be placed on interrelatedness. We are interrelated with one another in such a way that humans are part of the landscape rather than separate from it. This may seem like a subtle distinction, yet its implications are profound for our way of being in the world.

We often refer to the global environment, for example, yet this conception demonstrates some of the inconsistencies in our worldview. The word environment itself initially meant "that which surrounds," but it too has evolved to reflect the nature-culture dichotomy. Now, when we refer to the global environment, we envision the environment as it surrounds our earthly globe. Yet this perspective can be taken only by a disengaged humanity, perhaps by the omniscient gaze of objective science. This gaze dominates our understanding of knowledge, reflected in the dominance of sight-based metaphors for knowledge in this book. This gaze belies an ontology in which the environment is external to us rather than one in which we are embedded, a life world. The globe metaphor may thus contribute to our environmental alienation, in contrast to the sphere metaphor that typified the cosmology of earlier Europeans and indigenous cultures and which allowed them to live *within* the world. As an explicit example, consider how the metaphor of biodiversity hot spots (not to mention hot spots for endangered languages)— and maps of such hot spots—implies that conservation has to occur only in those special places on the Earth's surface, as opposed to everywhere.[39]

To learn to embody such interrelation, we might adopt one of the spiritual practices used to develop kinship with the world. We can imaginatively adopt the identity of an earthworm to feel lowliness versus that of an eagle to feel freedom and spaciousness. These are metaphoric projections, yet ones that allow us to connect with other beings. I am not the earthworm, at one level, yet perhaps at another I am. I am that. This is that, in metaphor. To empathize with the world, perhaps it is more important that we learn to effect such metaphors in our own experience, rather than invoking our habitual response: "That is an object," where "that" is another living being, there for our exploitation or experimentation. We would then embody a subjective relation to it, rather than an objectified one.[40] It is perhaps on such subjective and empathic relations, rather than facts alone, that a new and enduring conservation ethic can be built.

Other scholars have proposed specific metaphors that might reform the nature-culture duality and hence our relation with the planet. John Ehrenfeld, executive director of the International Society for Industrial Ecology, for example, has proposed industrial ecology as "a new paradigmatic metaphor" that will help us attain sustainability through novel solutions to environmental problems. He highlighted how this metaphor is normative and beneficial because it derives from the metaphor of ecology, from which spring ideas of connectedness and other ecological principles. It encourages us to think of our industries and their processes as ecological systems; for example, in recognizing the role of detritivores, we would think about recycling as fundamental to society rather than an add-on. As a second example, Daniel Philippon, a scholar of English and ecocriticism at the University of Minnesota, has promoted "island" as a new guiding metaphor to help break down the nature-culture separation. He reviewed how the theory of island biogeography has been used to compare the ecological functioning of oceanic islands to islands of remnant forest on the mainland, which both highlights ecological limits (for example, by allowing us to see "Earth Island" as a potential Easter Island) and reminds us of the importance of the connecting matrix between distinct islands.[41] Each of these overarching metaphors has appeal, but they all tend to emphasize a single metaphor rather than promoting playful inquiry with diverse metaphors.

Turning to the second meaning of interconnection, we seek metaphors that connect us to one another. Just as we must keep in mind that the extinction of relationships between species may be just as insidious as the extinction of species themselves, the neglect of human relations in the context of the problems we face may be just as debilitating as the problems themselves. For example, environmentalist concern for mountain gorillas led to eviction of the Batwa pygmies from the Bwindi Impenetrable Forest National Park in Uganda in 1991.[42] This was justified as a way to isolate the gorillas from human interference, motivated by a desire to maintain them as a crucial expression of biodiversity. But note that biodiversity here is separate from people and that the human agency actually responsible for gorilla decline was not questioned at all. It was assumed that humans have a malignant presence, always and everywhere, even though the Batwa had lived in that landscape for generations. Ironically, however, the people who remained were the environmental elite who had access, through specialized gear, to dehumanized landscapes with romanticized gorillas. It has since been realized that removing the Batwa was a mistake because their presence had reduced poaching and other harms to the gorillas. If policy discussions had focused on how humans should interact with the natural world—the appropriate type of agency and associated metaphors—then the Batwa might not have been removed. Metaphors engender particular practices, and we cannot ignore how these practices influence the relationship among peoples.

We also need to reconsider our metaphors because language can contribute to direct conflict between peoples. My first example relates to the previous one. Some African governments have enacted "shoot-on-sight" policies over the past several decades to protect biodiversity in national parks against poachers, an approach that has been interpreted through the lens of a militaristic planning metaphor. The problem is that not all poachers are created equal: they may in some cases include impoverished women looking for small game or fish to feed their children. Violence against them reinforces cyclic patterns of violence in these parks and rests on a problematic contrast between personified wild animals and animalized poachers. The defense of biodiversity here relies on an us-versus-them duality that encourages militaristic opposition and escalating violence, including human rights abuses.[43]

Sometimes such conflict occurs between different stakeholders involved in an environmental policy debate. Two interdisciplinary ecologists, for example, analyzed the language used by those arguing for culling versus translocation of hedgehogs from islands in the Outer Hebrides, Scotland. They found that the arguments of the two groups differed significantly; the former used scientific language to make a case for the benefits of a cull for local shorebirds, and the latter used more emotive language to emphasize the welfare of the hedgehogs. The media exacerbated this dichotomy, thereby fueling conflict between the groups and preventing them from exploring areas of common ground. In another example, two communications scholars demonstrated that the 1988 Yellowstone fire debate was coined rhetorically with two archetypal metaphors—death and rebirth—that cohered with competing worldviews for park management, an ecological-holistic one (fire causes rebirth) versus a human-centered one (Smokey the Bear: fire causes death). In this context the metaphors "invented" alternative responses to the situation: the fire was seen as a crisis by those who viewed fire as death, but the crisis was not so apparent to those who saw it as a necessary part of a natural cycle followed by rebirth. In both these examples, a discussion of underlying metaphors and some "rhetorical jujitsu" might have helped reduce conflict in policy formation.[44]

In summary, we require metaphors that connect people to both ecological systems and to other people. We cannot rely on stilted scientific language, but instead require resonant metaphors. We may need to promote our ideas, as discussed in the next chapter, though the deeper question remains: which values do we want to emphasize? It is this question I look at in the following two case studies, beginning with questions about the type of nature-humanity interaction encouraged by the metaphor and practice of DNA barcoding.

V

When Scientists Promote
DNA Barcoding and Consumerism

No generation is given the opportunity to evade the moral
ambiguities of its own intellectual enterprises and choices.
—*Anne Harrington,* Metaphoric Connections

Who assigns names and numbers
to the innumerable innocent?
—*Pablo Neruda, "LXIV,"* The Book of Questions

Humans share planet Earth with an astounding variety of species. Over the past few decades, we have come to relate to them with a predominant metaphor from the natural sciences, biodiversity, which came to rapid prominence starting in the mid-1980s to communicate the severity of species' losses.[1] One of the great challenges for conservation biology is to identify and categorize these species in the interests of sustaining them—and by association, maintaining the myriad benefits and services that derive from their presence. The Chilean Nobel laureate Pablo Neruda probably wasn't thinking of this issue when he penned his poem, yet his question is nonetheless pertinent to human attempts to classify biodiversity.

In what follows, I consider a tool for documenting biodiversity that has been promoted through a particular metaphor, DNA barcoding. It is a metaphor that arose not just from looking at biodiversity in nature, but from an observation in a supermarket. This metaphor thus links biodiversity with consumerism in a compelling way, demonstrating how metaphors and their values interweave modern science with deeply held ways of being in modern industrialized societies. The question, as Carol Kaesuk Yoon put it in *Naming Nature,* is whether "without even realizing it, we have traded a view of ourselves as living beings in a living world for a view of ourselves as consumers in a landscape of merchandise."[2] It is through this cultural lens that the proponents of DNA barcoding intend for people to observe and relate to biodiversity. I wish to consider whether this metaphor and the associated technology are really that likely to improve our connection with the natural world and its species. This case study thus moves us beyond questions about the effect of metaphors coined in the distant past to questions of scientific responsibility for those of the recent past and the present.

Let us first consider in more detail the creation of a new metaphor. If we imagine a scientist thinking about a phenomenon, how does she come up with a metaphor? It might be a creative insight or it might be a suggestion from a colleague. But is a scientist led inexorably to one metaphor, or does she winnow from a set? How does that process occur? If a metaphor works epistemically, does a scientist just adopt it no matter what? Is this appropriate? Or is there any possibility of stopping for a moment to think about its broader repercussions? For, as we have seen, once a metaphor becomes ingrained, it becomes very difficult to ponder it objectively. It becomes difficult to think of alternatives because we have named something in a particular way. It becomes part of how our world is constructed, and it tends to reproduce prevalent biases. Nonetheless, the fit between our chosen metaphors and the biological referents they describe is always loose (and referents may even depend on the metaphor for us to perceive them), and alternatives have considerable implications for how we relate with the world.

When a scientist coins a metaphor, she adopts it from a cultural source. There is no alternative—scientists simply apply everyday understanding within a specific realm that happens to be foreign to most people. Metaphor, as we know, means understanding one thing in terms of another. And that "other" that a scientist chooses will be something from everyday life that she shares with all of us. Thus, to select a successful metaphor she must draw on an implicit understanding of the larger linguistic ecology in which she is located. Her choice of one metaphor instead of another will reflect her social context as well as her ambitions. Noting that we describe the brain as a computer system, for example, David Edge, formerly a science studies scholar at Edinburgh University, recommended that we "explore the extent and the dynamics of this process by which our imagination comes to be dominated by those very devices which we devise in order to dominate and control our environment and human society."[3] When this happens we reinforce the way of perceiving instantiated in those metaphors, perhaps blinding ourselves to other ways of perceiving.

Because such metaphors derive from society, they connect with broader social associations and values, as we have seen. Metaphors are not merely shorthand for the facts; rather, they simplify complex reality by situating facts within a web of cultural meaning that gives them significance in our communication. It follows that the decision about what metaphor to use is not just epistemic. It is a rhetorical act. Scientists must explain their ideas so that others—whether their colleagues or the public—can understand and are thus attracted to those ideas and embrace them. Even though scientists may be able to bracket the social values implicit in a given metaphor in their practice (by operationalizing it, for example), they are nonetheless present. And with the help of these values, scientists can sell their ideas to both scientists *and* nonscientists; a well-chosen metaphor, because of its commonplace associations, unites the epistemological and rhetorical functions of metaphor and allows them to reach both audiences. Metaphors are chosen for their joint cognitive, emotive, and rhetorical benefits.[4]

A scientist might adopt a metaphor with its potential rhetorical or ideological significance in mind. As two American sociologists, Gary Fine and Kent Sandstrom, explained, "Metaphor . . . is a handy tool for the ideologist in presenting pictures of 'how things are' and of 'how they might ought to be'—pictures that both resonate with people's lived experience and offer them an appealing sense of how they can and should live. . . . Ideologies are characterized by the apparent abundance of metaphorical usage, not only in the conscious tropes that speakers employ, but also in their choices of images when they are 'just' communicating."[5] We never just communicate but always highlight with the metaphors that we choose, including those adopted in this book. Whether we view the environment as Mother Nature, mechanistic system, or renewable resource, genes as instructions or an ecosystem, or evolution as competitive or cooperative, for example, we emphasize certain aspects over others and thus paint a simplified picture of a more complex reality. When the environment is a resource we focus on its material uses, and when genes are instructions their developmental context becomes minimized and they appear to determine organismic outcomes.

Scientists may be relatively unaware of the broader associations of their metaphors because they originate within the encompassing cultural milieu in which they live. Consider the debate in the twentieth century between two influential American ecologists and their followers, Frederic Clements holding that integrated communities exist objectively in nature and Henry Gleason countering that communities are human constructs and that species are actually individualistic. When Michael Barbour, an ecologist at the University of California at Davis, interviewed eminent plant ecologists about the transition from a Clementsian to a Gleasonian view of vegetation in the 1950s, he was surprised to find that they saw little connection between this transition and large-scale, concurrent cultural trends toward individualism. He concluded that they had "remarkably little conscious curiosity about personal ethics or one's connection with society." Without such awareness, it will be difficult for scientists to understand the broader social context in which their research is situated. It may be even more difficult to question those constitutive metaphors that are part of the "hard core" of research programs discussed by some philosophers of science. These metaphors may form an integral part of what the Polish philosopher Ludwik Fleck called thought-collectives (or thought-styles), cohesive research groups formed by adherence to a particular, distinctive language.[6] To the extent that these groups are committed to their language, they may not realize how much the metaphors informing the core of their research agenda reproduce a particular social context.

Scientists are authorities within society, so when they promote a metaphor they effectively endorse its implications. The question is whether this occurs openly. A basic ideal of democratic systems is that there is free speech and open deliberation about alternatives in the marketplace of ideas. It is illegitimate to stifle such debate and discussion, because some views then remain unrepresented. Representation may be suppressed in various ways, however, particularly by those individuals who have power and influence that allow them to sway the debate in their direction, thereby reinforcing their privilege. For decades, social theorists have detected a scientification of politics, whereby scientific authority is increasingly used to ground decision making and thus obviate normal democratic processes.[7] In the domain of biodiversity, for example, environmental scientists are the locus of authority and looked to in various ways; their opinions are sought by the media, by committees dealing with environmental issues, and by environmental organizations. There is a particular risk that people may treat scientific metaphors as literally true, as science is thought to use the language of objectivity. In these and other ways, scientists can implicitly or explicitly promote a particular worldview with their metaphoric choices.

The "selfish gene" metaphor provides a fine example of the implications of such metaphoric choices. Richard Dawkins coined this metaphor in his best-selling book of the same name, which has become one of the founding texts for evolutionary psychologists. The survey I discussed in earlier chapters found that over 59 percent of evolutionary psychologists agreed or strongly agreed with the statement "Animals are competitive because of their selfish genes," whereas fewer than 34 percent of the respondents from the other organizations did. This suggests that it has become a constitutive metaphor for the former group. At the same time, I suspect that 72 percent of the evolutionaries disagreed with this statement because of its popular resonance. It has led many people, as it did them, to associate evolution with competitiveness and selfishness, regardless of whether "the shift to a metaphorical selfishness at the level of genes was precisely what allowed for real altruism at the level of organisms."[8] Dawkins may well have intended for this metaphor to politicize science in the interest of promoting his outspoken antireligious views.

Scientists have an enhanced capacity to promote their metaphors through the societal funding they receive for their research. Although scientists may intend to be objective and disinterested, their language displays an intersection with commercial and political interests. By using a metaphor, a scientist encourages the media to use it too, thereby promoting any associated ideals. This augments the scientist's privilege, not least because of a positive feedback whereby he can obtain further funding to continue building that perspective of the world. Funding thus elicits the performative power of a metaphor by allowing a scientist or group of scientists to begin to construct society in accord with it, affording even more power to those who participate on the ground floor of such developments. In the case at hand, for example, scientists might dictate how people will relate to biodiversity. Their metaphors are not merely words; they entail performative action. They fashion a particular view of the world, rather than another. As the historian James Bono of the University at Buffalo explained, "The work of metaphor . . . is not so much to represent features of the world, as to invite us to *act upon* the world *as if* it were configured in a specific way *like* that of some already known entity or process."[9] Certain outcomes become predestined. By using metaphors from everyday sources, biologists have the power to gradually fashion the natural world in their own image, both in thought and deed.

Scientists may intend well with their metaphoric choices, but they are under tremendous pressure, like any other advertiser, to sell their perspective. The American sociologist of science Dorothy Nelkin, drawing on a case study of the deterministic metaphors used to represent human genetics, showed how they sometimes do so with "promotional metaphors." She reviewed how scientists use "evocative images, catchy titles, and often corny metaphors" to attract an audience and suggested that they should restrain such tendencies to maintain public trust.[10] To the extent that an audience realizes that a given metaphor highlights certain aspects of an issue rather than others, its members might wonder about what is left disconcertingly out of sight. Even to assess such possibilities, we cannot assume that natural science leads inexorably toward sustainability progress. Though we have in place certain societal checks and balances on the ethical implications of research into new technologies, we need to extend them into the potentiality of particular metaphors. It is with metaphors that technologies begin.

DNA Barcoding and Consumerism

To examine such issues, I consider a case study from the realm of biodiversity science, DNA barcoding. A DNA barcode is a

short DNA sequence used to discriminate among species. DNA
has long been used to identify taxa, the innovation of DNA
barcoding being to standardize taxonomy by using a single
DNA sequence, and potentially the shortest one possible, to
differentiate taxa across a wide taxonomic range. It appears
that many organisms, aside from plants, can be discriminated
using a single barcode, the first 648 bases of the conserved mitochondrial gene COI.

DNA barcoding lends itself, analytically, to consideration
of promotional metaphors for a few reasons. It is first of all
extremely current, having initially appeared in 2003 in a paper
in *Proceedings of the Royal Society of London* written by the
Canadian biologist Paul Hebert, the father of DNA barcoding,
and three co-authors. This paper has already been cited over
one thousand times. The recent coinage of this metaphor means
that we are very close to observing the process of new metaphor
formation as it happens. We can also understand its current
usage more easily than might otherwise be the case, as it can
be traced back to a single scientist, Hebert, whom I was able
to interview regarding his adoption of the metaphor.[11] He recounted that he came up with the idea while walking in his local
supermarket one day; he began to wonder whether the barcodes
there might be applied to identifying species. In addition to its
epistemic aptness, Hebert explained the motivation for its use
in terms of a public pitch: "We were heading into a campaign
for serious public support for the enterprise. . . . The initial pitch
was really to the broad readership and not to members of our
scientific community. I would never have titled it 'DNA barcodes' if I were writing a paper for my five scientific peers. . . .
You want to be the flavor of the day." More trenchantly, Hebert
asked, "How do you present a revolution?" Notwithstanding my
earlier points that we need to query the social dimensions of

any scientific metaphor, this one was intentionally chosen for a readership broader than just other scientists. Its social resonance was to some extent given priority over its epistemic value. In part because of its promotional vein, the metaphor has not been without detractors. Numerous critics questioned whether the technology would work, particularly in differentiating species within young and diverse groups of species (clades). Some felt that the metaphor drove the investigation before its empirical adequacy had been assessed and that it was simply a way to obtain media attention, as Nelkin predicted. They pointed out, somewhat cynically, that the metaphor was not abandoned even after it was discovered that there is no universal barcode for all taxa. As the systematist Quentin Wheeler stated, "It is telling that the abandonment of the core tenet of barcoding did not distract proponents from their quest for funding."[12]

Such metaphoric promotion is mainly in the province of established senior scientists such as Hebert who have the influence to dally with resonant language. Hebert recounted problems with his scientific colleagues: "We took a lot of heat for using that term. . . . We had an early hit on our credibility." Some of the attacks were "quite violent," and he received "incredibly inflammatory e-mails." Thus, he acknowledged, "It would have been much more dangerous for a young academician to go forward with this metaphor. I don't think a young academic would have had the allies, so they wouldn't have survived the heat." Such heat is felt only by a scientist who is in the business of promotion.

In short, DNA barcoding is an exemplary promotional metaphor because it sought to promote a vision, namely, that it would contribute to the conservation of biodiversity. On these grounds, its proponents would probably argue that the severity and invisibility (to most people) of biodiversity loss make it imperative that biologists do whatever they need to do to reduce this trend, which includes framing it in whatever way they feel will obtain the funds they require. As a conservation biologist and natural historian, I am sympathetic, yet my overall response is perhaps best expressed by David Takacs, in his review of ideas of biodiversity: "How can I balance my healthy skepticism about conservation biologists' proselytizing on behalf of biodiversity against my fervent hope that they succeed?"[13] We need to inquire critically into whether the promotion of DNA barcoding is in the best interests of biodiversity conservation.

To answer this question, we must in part assess a particular entrepreneurial technology through which DNA barcoding promises to conserve biodiversity. Specifically, a guiding image has been the Star Trek tricorder, the hope being that we can someday identify species just by putting a piece of them in a small handheld machine, the Life Barcoder (Figure 10). The Life Barcoder could also present encyclopedic information on each species from the Internet. It would open up an array of practical benefits, including the ability to identify organisms on the basis of small fragments, dead or alive, such as fish in a market, old feathers from dead birds, or potentially invasive species at borders. The estimated cost of developing such a machine is between one and two billion dollars.[14] It is the grandiosity of this vision that has drawn extensive media coverage, and the metaphor of a barcode has provided an image that people can grasp.

It might seem that barcoding was essentially going on before the metaphor was coined, that taxonomists were already doing this anyway, though by another name. But this new name is crucial here. The metaphor did something: it has made all the difference by creating a unified cause. According to Hebert, the

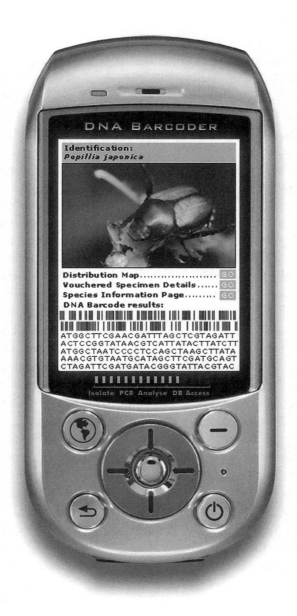

Figure 10. An artist's impression of a Life Barcoder.
Image courtesy of Robert Dooh.

metaphor has played a significant role in obtaining tens of millions of dollars in funding. Like metaphors used in the Human Genome Project, such funding contributes to a near-religious vision and sense of meaning to people working on the project.[15]

You might contend that I am confusing the metaphor with the technology. The two coincide, however, because the technology could not be developed as it is without the metaphor. A technological metaphor is a halfway point between dream and reality. It spreads in advance of the technology itself, providing fertile ground in social support. It may appear that the actual work has to do with sequencing a mitochondrial gene ("that's what they're *really* doing"), yet the lines observed in molecular assays are only comparable to barcodes; hence the metaphor. By this act of comparison, we set these taxonomic questions in the context of how we look at items on the shelf in a supermarket. This comparison helps sell the technological development. Thus, the metaphor is not just rhetorical embellishment; instead, it drives a particular vision of the world and how it *should* be. The metaphor is not just language but an encouragement to people to act on the world in a particular way and to develop certain capacities rather than others.

The barcode metaphor reflects a cultural moment when consumerism is prevalent. The barcoding system used to facilitate purchasing has thus become quickly constitutive of thinking about biodiversity. When I interviewed Hebert in his office at the University of Guelph, he stated that barcoding is "less of a metaphor than the Tree of Life . . . in that it relates directly to the major product lines of life—species." The Tree of Life is a major metaphor used to describe evolutionary history, and although Hebert stated, "I'm not sure any phylogenetic relationship looks much like an oak tree," he did not extend similar logic to whether species are really like product lines.[16]

Rather than Darwin's organic metaphor, rooted in experience in nature, we have a replacement that reflects humanity alone. Species have become product lines.

The metaphor thus promotes a neoliberal capitalist way of observing species. This is an easy metaphoric extension for people, and it facilitates their understanding, because barcodes are such a normal part of daily life in modern Western society. For this reason, its contextual values remain somewhat hidden. As a feedback metaphor, it nonetheless integrates prevalent cultural values into scientific research. We have already seen that the metaphor was thrown intentionally into the social sphere with particular societal and political outcomes in mind. It reflects a consumeristic way of organizing and relating, which exemplifies how metaphoric values may influence our way of thinking, our science, and thus our approach to biodiversity. Its proponents have put themselves squarely in the realm of social and moral critique.

Like much consumerism, DNA barcoding envisions that the solution to a problem lies in a product, in obtaining something that we currently lack. It needs to be marketed to fulfill a need. Among other benefits of DNA barcoding, mainly for scientific inventory of biodiversity, the Consortium for the Barcode of Life lists two broad benefits that fulfill just such needs. First, the Consortium claims that DNA barcoding democratizes access to biodiversity by "empower[ing] many more people to call by name the species around them." Second, the Consortium claims that DNA barcoding will "engender appreciation of biodiversity both locally and globally."[17] Both of these propose that the net benefit of DNA barcoding is that it would connect people with the natural world, but is that likely to be the case? I contend instead that DNA barcoding only weakly supports sustainability as an objective because it strengthens the dominant worldview rather than aiming to question it more deeply. It is fundamentally conservative, reinforcing the consumeristic way that we view the world. At the very least we need engaged, public discussion about whether this is the case earlier in the development of new technologies such as DNA barcoding, rather than later on.

An Obligatory Passage Point

Just as barcodes allow us to standardize products and to track their movements, species would be standardized and trackable with DNA barcoding. In this way, the metaphor of a barcode might suggest that DNA barcoding will be a democratizing force. That is, it would increase connection with biodiversity by allowing more people access to it. As CBC News reported, "The findings open the door to a radical change in how animal species are defined and identified, and could change how amateur ornithologists and other nature lovers observe the world around them. Hebert envisions the creation of hand-held devices that would allow the average person to identify plants and animals within minutes by analyzing their genetic bar code." This is one possibility, yet there are others that remain hidden by this metaphor. Just as Geographic Information System technology was promoted as a way to democratize mapmaking capacity, that is only a potentiality, one that can be overridden by the tendency to attach truth to the interpretations that flow from it and thus power to those who control them.[18]

People may one day *require* a barcoder to be sure of the identity of organisms around them. In certain contexts this democratizes such knowledge, but in others we would no longer be able to rely on the old-fashioned ability to identify organisms in the field with our own senses. DNA barcoding has detected

Figure 11. A potential new cryptic species, owing to DNA
barcoding, the common raven. Photo courtesy of James Kamstra.

numerous pairs of cryptic, or look-alike, species—for example, within common North American birds such as the hermit
thrush, western screech owl, and common raven (Figure 11).[19]
Though molecular techniques detected cryptic species previously, the genetic markers they used were more often associated
with traditional characters in part because there was no dream
of ever being able to apply such technologies outside the lab. In
contrast, to understand the potential implications of splitting
these species in two, imagine that one species of common raven
is widespread whereas the other is rare, and that there is no
way to distinguish between them except through DNA barcoding. This certainly would be a radical change: you would need
to catch ravens to take a feather from them to identify them,
rather than just being able to observe them in the field. As an

extreme possibility, users would need to put every individual through their Life Barcoder in case visually similar individuals represented different species. Naturalists' capacity to identify species in other ways, using their own senses, would probably atrophy. Rather than the emphasis being on human-scale observations and interactions with the natural world, it would be on the technology.

The metaphor also obscures how, like any consumer product, the Life Barcoder would be too expensive for most of the world's people. Keep in mind that the majority survive on only a few dollars per day. Given the history of technology, it seems unlikely that barcoders will eventually be widely affordable, let alone the free gadgets that at least one biologist has predicted.[20] It is just as likely that DNA barcoding would restrict access to biodiversity to relatively wealthy organizations (and possibly individuals), who could afford the requisite technology. You might counter by referring to the expense of taxonomic manuals—not to mention their unwieldiness—yet people can learn from them and apply this knowledge to commonly encountered species, which, as we have just seen, would in at least some instances be less likely with a Life Barcoder.

DNA barcoding might democratize access to biodiversity for certain people, then, but it is much less clear that it would do so for others. There is widespread scientific support for DNA barcoding around the world, instantiated in the Consortium for the Barcode of Life, yet it appears that it is being developed with little input from, or collaboration with, those who deal with conservation in particular places, on the ground. There is considerable agreement in the literature that bottom-up participatory conservation with local people is more effective than top-down approaches. This mirrors arguments for earlier public engagement in the development of new technologies, which include those that are normative (engagement is beneficial for including stakeholder values), instrumental (engagement increases legitimacy of and trust in decision making), and substantive (engagement leads to better decisions).[21] Each of these provides reasons for including a broader range of people in decision making about the development of technologies such as DNA barcoding.

On the basis of ethnoecological research at Mt. Kasigau, a biodiversity hot spot in Kenya, for example, two geographers, Kimberly Medley and Humphrey Kalibo, concluded, "A conservation agenda that asks local residents to accept the global importance of biological diversity in a region ('Think Globally') and then to act as a participant toward its protection ('Act Locally') will not promote collaboration in resource protection."[22] The problem is that this presumes the value of biodiversity as understood by Western science and thus devalues the contribution of local citizens' knowledge to conservation initiatives.

As far as I can tell, discussions about DNA barcoding have ignored whether its purported democratization would extend to indigenous ways of knowing biodiversity. Certainly, indigenous peoples might adopt this technology, as they have done with so many others, perhaps adding their traditional knowledge to the database provided in the Life Barcoder. There is a risk, however, of embedding problematic assumptions. Although some indigenous peoples have not safeguarded local landscapes, others continue to live in close relation to the land, as they have for millennia, and they have developed precise, practical taxonomic systems that contribute to their interaction with local biodiversity. Even more significant in the current context is the fact that they often speak endemic languages that contain entire systems of understanding that contrast with ours. These peoples and their languages are disappearing fast, and as we lose them, we lose alternative ways of knowing, including ways of living on the land.

Now consider barcoding. It is promoted as the universal language for biodiversity conservation, as the true way of seeing, which contrasts markedly with the intimate indigenous stories about local landscapes and biodiversity.[23] The names on a barcoder would not connect with living beings in the same way as indigenous understanding. DNA barcoding could contribute to the stripping of those connections because it seeks a uniform story using COI rather than listening to the diverse ways people have related to and classified species in local contexts. It is also likely that DNA barcoding would become institutionalized, especially given the tight linkages among biodiversity science, commercial interests, and government agencies. This would further threaten indigenous peoples' ways of knowing the species around them—a critical component of their world— because it works top-down and assumes we need this scientific biodiversity knowledge rather than working with what people already know. It assumes, as in the case of the Batwa described in the previous chapter, that the answer is in a scientific key rather than in a diverse harmony. DNA barcoding could thereby contribute to the loss of other languages and knowledge systems in use.

To understand the consequences of this metaphor and its associated technology, consider an analogy. In his book *Shadows of Consumption,* the political scientist Peter Dauvergne explored the environmental consequences of consumption drawing on five case studies: automobiles, gasoline, refrigerators, beef, and—being a Canadian—harp seals.[24] He demonstrated how the costs of new technologies increasingly spill into distant places and times, further displaced from the consumer. Metaphorically, these are the shadows of consumption. These shadows tend to fall onto those with less power, whether indigenous people in the Arctic, citizens of the developing world, or future generations. Consequently, he suggested that wealthy policy makers, from whom we most often hear, are often unduly optimistic about the environment because those absorbing disproportionate costs cannot protest. In the promotion of barcoding, we see a similar process whereby the potential shadows of this metaphor, consumerism being one of them, are swept aside by those promoting it. They can be optimistic about its benefits, whereas those living on the ground may be faced with the challenging decision of whether to adopt it in a local context amid much greater issues of survival.

More broadly, once DNA barcoding objectifies species as commodities, we require managers to organize the information about them, a further way in which what was once distributed knowledge becomes restricted to very few individuals. It appears likely that DNA barcoding will become what science studies scholars refer to as an obligatory passage point.[25] An obligatory passage point is a claim, an article, or a technology that people—whether scientists or the public—must accept or adopt to be able to reach their objectives. In the realm of science, this garners even more support for a certain line of research. Thus, those developing DNA barcoding obtain tremendous power for organizing our view of biodiversity and our approach to it. Just as agricultural companies can program the social organization of food production through their products, a barcoder would allow its proponents to program the social organization of our approach to biodiversity. Life Barcoders might themselves be distributed, but the structure of the knowledge they contain would be determined elsewhere.

Stated another way, there is a compelling circularity in DNA barcoding because its proponents simultaneously define what biodiversity is and provide the only way to measure it. A similar process has occurred in the way that molecular biologists have defined life itself through their technologies. Although this might be acceptable if DNA barcoding were the correct way to identify species, the technology does not avoid intractable questions about how to define species and which species concept we should use. Yet by fixing a certain species concept within a technology, DNA barcoding overrides such questions and solidifies the elitist role of a particular strain of biodiversity science. As two eminent science studies scholars, Geoffrey Bowker and Susan Leigh Star, observed in their book on the consequences of classification, *Sorting Things Out:* "Each standard and each category valorizes some point of view and silences another. This is not inherently a bad thing—indeed it is inescapable. But it *is* an ethical choice, and as such it is dangerous—not bad, but dangerous."[26] This is all the more reason to ensure that the very decisions about whether to pursue such choices are themselves distributed and made democratically, rather than by scientists alone.

Those who control DNA barcoding could obtain the technological funnel through which biodiversity and its status would be assessed in the future. In terms of the metaphor *spaceship Earth,* lead scientists become the pilots who would control our direction and outcomes. DNA barcoding provides the final managerial instrument for biodiversity conservation because it seeks to resolve the problem of taxonomy so that all species can be lined up on shelves for our use and managed as product lines of life. This vision reveals faith in a technocratic system, one that seeks top-down, centralized control and solutions for biodiversity management. It epitomizes the belief that we need to identify all the biodiversity pieces to know how to conserve them. This belief may be understood as part of the "almost religious conviction that a widespread adoption of computers and communications systems along with easy access to electronic information will automatically produce a better world for human living." This ideology originated with the information technologies developed for the Cold War to provide global technological oversight and has since spread widely through modern society. Proponents simply assume that DNA barcoding can be properly extended to provide this oversight to biodiversity and its conservation. In actuality, it may just exaggerate the problem of taxonomic inflation, having more and more species that we don't know what to do with. It is a progressive vision rooted in a someday technological utopia where everything will be figured out, and it appears that this vision, rather than a deeper intention to democratize access to biodiversity, is what drives it.[27]

Caring for Biodiversity

Just as we appreciate many of the products marked by commercial barcodes, the metaphor of DNA barcoding suggests that the technology will not only connect people to biodiversity by allowing more of us to access it, but also deepen our appreciation for it. This assumes, however, that we require more scientific knowledge to inspire us to care for biodiversity when in fact there is ample evidence that we already do. The impediments to acting on this belief are not so much taxonomic and technological as they are economic and political.

Furthermore, there is little evidence that simply putting a name on an organism leads one to care for it. Rather than caring being a consequence of naming, perhaps one must care about nature enough to name species at all. Naming may therefore merely reinforce a preexistent sense of caring. The lives of distinguished conservation biologists suggest that this is the case. For example, the esteemed Harvard biologist and conservationist E. O. Wilson grew up as a naturalist, and he stated: "Hands-on experience at the critical time, not systematic knowledge, is what counts in the making of a naturalist. Better to be an untutored savage for a while, not to know the names or anatomical detail. Better to spend long stretches of time just searching and dreaming."[28] It may be that the quest to find new species and to identify them inspired many biologists to choose their careers, and that access to an easily available label might have undermined and even squelched their exploration.

People develop an appreciation for life primarily through interactive experiences with organisms, and it is unclear whether the Life Barcoder would facilitate these interactions. We desperately need people who have developed an intimate relationship with the natural world, including its organisms, and naming fulfills only a fraction of this need. Consequently, we must train the next generation of naturalists, rather than kids with Barcoders in hand. By looking closely at an organism, people learn more than just its name: they learn how to observe—a very valuable scientific skill in its own right—and they learn something about this particular individual, this species. Over time, they learn about the concept of variation, which is so critical to understanding biology yet potentially veiled by the implicit essentialism of DNA barcoding. It is also unclear how this technology would increase appreciation for biodiversity, given that it means interacting with a screen rather than with organisms themselves. Recent studies have shown a significant correlation between the decline in visitation rates to national parks in the United States and videophilia, "the new human tendency to focus on sedentary activities involving electronic media."[29] We simply don't know how Life Barcoders will affect this trend.

It is debatable whether DNA barcoding would increase the value of other species or devalue them. Yet Hebert opined, "I actually would not mind the fact of humanity starting to think of species as important items on the store shelf of life. I mean, I really think that might be a very progressive step from the current view, which is that you can't read these things at all."[30] This statement instantiates the stealth issue advocacy of the metaphor by emphasizing species' instrumental value, their usefulness to human beings. Some scholars would echo that this is one of the strongest arguments for conserving biodiversity, potentially aligning biodiversity and economics, as we see happening in the trend toward green business. In contrast, others declare that if we reduce biodiversity to information with only instrumental value, we undermine some of the more profound reasons for conserving species. A utilitarian perspective is not the only way to encourage people to care about other beings. If species are valuable only if they are useful, then there is little justification for conserving the remainder that aren't. Through its association with consumerism, DNA barcoding would tend to emphasize knowledge about species that serve a purpose for us—which we can already see in the priority given to developing barcodes for commercially important species such as edible fish. This debate is ongoing; the question asked here is whether it is appropriate for scientists to sell their personal answer to it with a metaphoric choice.

This discussion touches on two other prevalent feedback metaphors used by environmental scientists in the interests of sustainability: ecosystem services and natural capital. Their basic insight is that for human activities to be sustainable they must maintain natural capital, the stock of assets that ecosystems require to perform functions such as carbon sequestration, pollination, and water provisioning. Natural capital is a

powerful metaphor in its blending of nature and culture. As
the Finnish environmental policy scholar Maria Åkerman observed, however, "The polysemy of the metaphor of natural
capital made it possible to underline different connotations
and attributes of words 'natural' and 'capital' on different oc-
casions." This has contributed to a struggle between ecologists
and economists to determine whose disciplinary meaning will
prevail and thus who will obtain the power to lead us into a
sustainable future, a general tendency being that the metaphor
"downplays meanings of nature and natural other than those
that are related to production."[31]

Natural capital may assist us in recognizing the ways humans are embedded in and dependent on the natural world,
rather than thinking of conservation as something that happens in wilderness areas set aside from human involvement.
Yet it might subsume conservation under economics. It may
thereby contribute to the capitalization of nature—conservation planning being done in terms of "investing" and "portfo-
lios" is a good contemporary example. Accordingly, scholars
continue to debate whether it is appropriate or helpful, and
specifically whether it implies a weak sustainability, where man-
made capital might substitute for natural capital, as opposed to
a strong sustainability, where these forms are complementary
and at least some natural capital is inviolate. Norton argued,
for example, that natural capital is an inadequate basis for sustainability because "specifying stuff that, if lost, will negatively
affect welfare doesn't go far enough: if we save just the stuff that
has a measurable impact on welfare, it is welfare that counts,
not stuff." This is not to deny that the metaphor is useful in
certain contexts, but it might contribute to forgetting that "the
model of nature as capital comes at the price of downplaying
or misrepresenting some key respects in which our relation to

it is radically *unlike* our relation to any capital stock." In this respect, perhaps the metaphor of ecosystem services, although still utilitarian-oriented and directed to ecomanagerial solutions from Western science, might be preferable.[32]

I do not wish to imply naively that we might escape a consumeristic worldview, yet we nonetheless need to be aware of when we might be implicitly taking consumerism as a good in its own right. There are other values in life besides whether we can purchase something; some would say that the greatest beauty, for example, lies in things that we cannot own in any sense, such as a sunset or recently fallen snow. As Paul Evans observed in his article "Biodiversity: Nature for Nerds?" the discourse of biodiversity tends to put nature in the utilitarian part of the equation, so that all kinds of other things we care about are left out. The result is a landscape "determined by technocrats driven by big business."[33] Barcoding exemplifies this tendency to attempt to quantify everything in a consumeristic vein.

By accentuating species' instrumental value, DNA barcoding may simultaneously objectify them. The cover of the December 4, 2004, issue of *Science News*, which had an article on DNA barcoding, carried a striking image of animals with barcodes on their foreheads (Figure 12). The animals and the codes blend and become one, turning biodiversity into an array of commodities. Although this image is clearly meant to be arresting and suggestive, DNA barcoding might well reduce species into objects able to fit within the human accounting system. This would heighten the tendency to regard everything in our world as a consumer product. It is as the German philosopher Martin Heidegger observed: once we treat forest trees as a commodity, they become nothing but a standing reserve, there for our use.[34] Yet this very tendency to give priority to commodities contributes significantly to the loss of biodiversity.

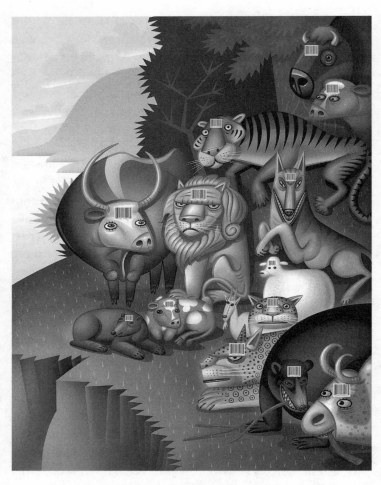

Figure 12. This image, which appeared on the cover of *Science News* magazine, insinuates how barcoding converts biodiversity into an array of commodities. Image used with permission of Dean MacAdam.

Conservation might flourish if we instead subjectified other species, recognizing them as independent living beings with entire life worlds of their own. They may be lines of life, but are they products for our use? Some of the products we purchase are cheap kitsch that gets thrown away after a few days—presumably not how we should treat the species that are of concern to conservationists.

It is through objectification that barcoding resonates in more ominous ways. If this metaphor for biodiversity becomes fully naturalized, what would be the next step? Hebert initially felt that the barcoding metaphor was "beautiful" and "just perfect," and though it had "serious baggage in the scientific community . . . it resonated immediately with the public. It drew in the unwashed masses." Upon reflection, though, he also acknowledged that it contributed to various misinterpretations. Would species—or even human individuals—eventually carry barcodes on their foreheads? Some people associated DNA barcoding with Craig Venter's wish "to provide a whole genome sequence for every human being at birth for a thousand dollars," to which Hebert replied, "We're so far from that it's not even funny." Hebert incisively discounted such possibilities as "bizarre Orwellian ideas." But are these ideas bizarre and Orwellian? Given the web of meanings of this terminology, I would suggest not. Although this technology itself would not apply to humans because we're all members of one species, in a sense the control it seeks over other species is quite Orwellian. We could be the Big Brothers of biodiversity. And we know from human history that technologies are extended in unforeseen directions, so we cannot exclude that possibility for DNA barcoding.

Once species are commodities, we can do with them as we wish. Hidden at the heart of DNA barcoding is the fact that in some cases we would need to maim or kill species to identify them—a convention within much of normal taxonomy too, but one sometimes exacerbated by DNA barcoding. This contrasts markedly with how previous cultures valued real, living individuals too much to kill them in the name of an abstraction such as biodiversity, doing so only as a matter of survival. In modern molecular biodiversity science, however, such a view is merely quaint. In extending such an ethic, DNA barcoding might contribute to the depreciation of living beings.

A Better Future?

Every metaphor both highlights and hides, and thus, as we might expect, it has both strengths and weaknesses. We cannot subject these to a cost-benefit analysis, but instead must weigh features about which there is quite a lot of uncertainty. The previous sections, for example, have explored dimensions of the question of whether DNA barcoding will increase our connection with biodiversity. It is certainly possible that the costs I consider are outweighed by the proposed benefits, yet it nonetheless will behoove us to air such questions. Otherwise, we are left with an airbrushed image of its potential. By analyzing DNA barcoding more critically and publicly, we may better approximate the interests of future generations in its development, this being a fundamental element of most definitions of sustainability. The metaphor may seem to be just a label, but it performs an entire future for us, one that could not be created unless spoken, something that only a metaphor—and not a technology alone—can do.

I have shown some ways that DNA barcoding draws on its association with consumerism as a selling point despite the environmental costs of promoting the idea that more is always better. We cannot assume, in particular, that more access for more people to more species is equivalent to fewer yet higher-quality interactions by fewer people with fewer species. We want long-term, real outcomes for sustainability rather than just assumed ones. It is these interactions that we fundamentally care about, for people will act to preserve only what they care about. Given the gravity of the matter, we cannot just take it for granted that DNA barcoding will increase connection with nature—regardless of the allure of a new gadget.

In promising such a device, DNA barcoding might distract us from its framing of conservation as largely a biological and technological problem. Despite its aspirations, however, the Life Barcoder would not necessarily lead to the conservation outcomes we desire. Daniel Sarewitz, co-director of the Consortium for Science, Policy and Outcomes at Arizona State University, captured the general problem with the following question: "If humanity is unable or unwilling to make wise use of existing technical knowledge . . . is there any reason to believe that new knowledge will succeed where old knowledge has failed?"[35] The program to develop DNA barcodes implies that, with more information, we would short-circuit the myriad social problems thwarting conservation, including such critical issues as political and social inequality. Biodiversity science is part of the answer, yet we need to attend even more to causal, human elements grounded in the particular instances of people's lives on the planet. By framing the solution as technological, we may neglect alternatives that are simpler and more humanistic. As often happens, an ecomanagerial emphasis reflects the needs of the world's haves, whereas from the perspective of the world's have-nots the greater problem is simply inequality. Information in itself does not always empower people, and there are other ways to become empowered, including the development of solidarity through community and even successfully muddling through.

The metaphor also helps sell the technology by skipping the public inquiry and consultation that we expect in democratic societies. Hebert proclaimed, "We're repositioning humanity's relationship with life,"[36] but the question is whether this is something scientists should be doing. There has been very little ethical discussion of whether this consumerist vision is the institutionalized interaction we want to have with biodiversity, other than that among funded scientists and granting agencies. Thus, a particular way of approaching biodiversity is being advanced through richly funded channels and with very little public deliberation. Although a ripe and suggestive metaphor such as DNA barcoding has its benefits, it is the underlying values that need to be acknowledged and debated. By explicitly drawing on a prevalent cultural metaphor of this era, barcoding essentially writes us into nature.

There has been little public inquiry into the new future instantiated in this technology. Proponents of DNA barcoding have compared its novelty to the inventions of the printing press and the airplane: "The 1903 Wright Flyer was surely awkward, but powered flight soon proved scalable to a planetary level. In a similar fashion, a few studies . . . have now revealed the possibility that DNA barcodes will enable the identification of life to soar."[37] Such a narrative effectively tars as Luddites those critics who oppose technological invention. Instead, critiques such as mine merely ask of this progress the question "Relative to what?" "For whom?" and, I would add, "Over what spatial and temporal scale?" We now understand that airplanes have not only brought about benefits, but also contributed disproportionately to global warming. They are also unavailable to most of the world's people. Similar uncertainties apply to

DNA barcoding, yet they are glossed over here in the interest of promotion.

DNA barcoding also harbors a progressivist ideology. Just as we have rushed into other technologies without foreseeing their implications, we are now rushing ahead into this one. We have a problem, and more technology is the solution, regardless of either the cost or the evidence that more technology has often failed to solve our problems. One benefit of critical inquiry into our metaphors is to engage with such deeper questions. We might query whether we need to get off the escalator of progress at some point: when is enough enough? When do we stop—once all species are barcoded and all individuals are electronically tagged so that we can monitor their every movement? We assume that with more knowledge, we can finally figure things out. Instead, perhaps we need to temper our science too, just as we will need to temper our individual acquisitiveness and ask, "How much is enough?"

It may seem that such questions impede science too severely. Science is supposed to be a free form of inquiry, without constraint. If a scientist sees a similarity, and thus uses a metaphor, he should not have to look over his shoulder to consider what its implications might be. Certain risks have to be taken. By constraining science, we would lose opportunities for new discoveries and increased understanding that could make the world a better place. Even scholars such as Funtowicz and Ravetz, who seek a more socially involved, postnormal form of science, have raised concerns that a precautionary approach to scientific inquiry could be too restrictive. That could be the case, but as I will discuss further in chapter 7, the primary objective here is to create greater engagement between science and society over proposed metaphors. In addition, I hope I have already shown that we need to ask ethical questions about even purportedly pure science, though in the current context the science is applied from the start. We need to embody humility rather than hubris in the development of new technologies.[38]

If I am going to question the barcoding metaphor, it may seem that I need to propose a replacement. Otherwise, my case here may appear to be pedantic or mere sophistry: it is easier to critique than to create. I instead suggest that the creativity involved in adopting DNA barcoding as a metaphor was limited because it merely adopted the dominant cultural paradigm. When scientists are taken in by a particular paradigm, they merely commit themselves to the puzzle solving that constitutes normal science. Such normal science depends in part on entrenched linguistic resources; as stated by the historian Robert Young, "Paradigms . . . are only a scientized form of metaphor and are consequently more acceptable to the scientific community."[39] We thus see the need for critique of what might otherwise be seen as commonsensical. One of my prime purposes here is to recognize some of the limitations with what has come before because this is the first step in creating something new. Although I cannot personally call up an alternative that will work in all places and all times, I instead encourage all of us to become engaged in such issues. Together, we can better assess both current and proposed metaphors and technologies.

There were alternatives to the metaphor of DNA barcoding that involved less culturally loaded language. Hebert, however, discounted one such possibility, "species-specific DNA sequence tags," because he was "pretty sure the public would have rolled their eyes up and gone to sleep." He remarked, "If we had presented it gently, nothing would have happened. . . . To me, it's a big success, so therefore we can't have screwed up too badly." The rejection of this muted expression reveals the extent to which modern science is in the business of fund-raising and popularity, with all the related issues of how this might affect objectivity. Overall, however, I am questioning not so much the use of a resonant metaphor but the particular values that it highlights.

Another genre of alternative might clarify how the DNA barcoding metaphor affects our interaction with biodiversity. That is, with barcoding we relate to biodiversity as we do to food in a large-scale supermarket, whereas we might instead approach it as we would food in a local organic farmer's market.[40] Instead of viewing biodiversity as items on a shelf for our use, we are more in touch with the origin of what is before us. Here items are not so much laid out in a huge array of uniform individuals, but instead valued for the local context in which they have been nurtured. As buyers, we relate to these products as individuals, complete with their foibles. Buyers also interact more closely with the producer. Which metaphors for biodiversity might encourage this type of relation?

Finally, DNA barcoding might divert us from a very different emphasis for biodiversity conservation. If we want to get people out and connected with nature, why not just do so directly?[41] It is people's experience in nature that matters most of all. We could encourage such experiences for a fraction of the cost of DNA barcoding with widespread training of naturalists, thereby encouraging appreciation of nature. By focusing on naturalist training, we could sidestep the question of whether barcoders will connect people with nature. In this pluralistic vision, however, there is still a place for DNA barcoding, but decisions about it would occur as part of broader discussions about the appropriate way to transform our relationship with biodiversity. It is a concern that such discussions have happened so rarely in the case of DNA barcoding. And on this note, I turn to the next chapter, on how scientists incite fear to increase concern about invasive species rather than seeking to improve dialogue with everyday citizens who harbor differing views of biodiversity and its future.

VI

Advocating with Fear
At War against Invasive Species

*"The public" can engage in critical reflection about nature
and still conclude that the things they find of value there are
different from those prescribed by the culture of biological
diversity. Conversely, those advocating biological diversity
can do so thoughtlessly.*
—Kate Rawles, *"Biological Diversity and Conservation Policy"*

*Americans routinely use war metaphors whenever they want
dramatic change or want to stimulate public fervor. . . .
Undoubtedly psychological truths are concealed in this societal
penchant for approaching problems so belligerently.*
—J. W. Ross, *"The Militarization of Disease"*

I nvasive species are species that humans have introduced to a new geographical location, where they spread and may thus have various ecological and economic effects. They have become a prevalent theme in recent conservation biology. As a biologist, I understand the concerns about them, having spent numerous summers documenting their presence in significant conservation areas around the province of Ontario. I am, however, also skeptical because of how these species have been vilified, in part by scientific discourse. Such discourse is value-laden, and it distracts us from the conversations I think we need to be having. With growing global trade we can expect more invasive species, not fewer, so we will gradually need to reform our relation to them. This will require that we listen not only to scientific voices, but also to the many others who would have a say in decisions about them.

I contend that environmental scientists have been using metaphors that contribute to a climate of fear toward invasive species. They have not shied away from advocating on behalf of native species and against invasive ones, despite recent concerns in the scientific community about whether such advocacy is appropriate.[1] The question is whether a guise of objectivity is being used to buttress advocacy and whether this strategy is consistent with enduring socioecological sustainability. It may be that we need to reform the relation between science and society as much as the one between society and invasive species. This case study explores how our metaphors influence our relations not just with the world, but also with one another. This is important because we all—not just scientists—have a role to play.

Scientific advocacy against invasive species begins with the constitutive term *invasion* itself. Invasion is not a neutral metaphor, but a nationalistic feedback metaphor that raises associated fears that our country will be invaded by foreigners and our bodies by disease. One historian has claimed that the use of the invasion metaphor in the field of invasion biology derives from political geography, and ecologists have opined that initial concerns about invasive species arose from related concerns about Nazi invasion. Some scientists have pointed out that these cultural associations also confound scientific inquiry; the term *invasion* is misleading because it conflates spread with impact, when preliminary data suggest that species that spread are no more likely to have a significant influence than those that expand very little.[2]

In drawing on these cultural associations, invasion is an exemplary performative metaphor because we have difficulty conceptualizing invaders without immediately wanting to do something about them. As the philosopher Ludwig Wittgenstein stated in his *Philosophical Investigations*, "A *picture* held us captive. And we could not get outside it, for it lay in our language and language seemed to repeat it to us inexorably." We can't just sit by and let invaders spread. Put another way, who ever liked an invader? Thus, who would oppose someone who wishes to curtail an invasion? In this way, *invasion* goads us on. It is a wonderful example of the power of framing, analogous to the difficulty we have with opposing tax cuts, even if it is in our best interest to do so. Fears of invasion also reinforce the emphasis on the boundaries that separate us from one another. As the English scholar Laura Otis observed in her book *Membranes,* a study of the interaction between scientific accounts of and popular writings about the cell in the nineteenth century, "It is [now] fundamentally illogical to define oneself with borders in a world in which everyone is connected." Nonetheless, invasion biology continues to invoke such borders and boundaries, even though the movement of species continually overwhelms them.[3]

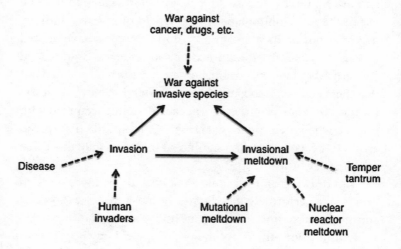

Figure 13. A portion of the metaphoric web of invasion and militarism associated with the field of invasion biology. The dashed lines indicate cultural resonance of terms and approaches within the field, as discussed in the text. Fear of invasion contributes to fear of a meltdown, and both of these encourage a militaristic response to invasive species.

I will not dwell on invasion itself much further in this chapter, instead focusing on two metaphors that flow from it and their implications. The first, *invasional meltdown,* more clearly reveals how metaphors in the field are meant to inculcate fear. And the second, *militarism,* demonstrates the response instilled by such fear. Together, these linguistic interactions create a metaphoric web by which we inculcate fear of invasive species and take action against them (Figure 13). I will discuss some of the ways that this approach is limiting; as in the case of DNA barcoding, we will see that the benefits of these metaphors are mainly assumed. More to the point, they do not cultivate interpersonal relations but instead use fear to reinforce hierarchical social relations. I will conclude by discussing alternatives, first by briefly considering the neutral approach advocated by some invasion biologists and then turning to richer alternatives that enable us to become more aware of what is obscured by the prevailing view of these species.

Invasional Meltdown: Advocacy by Fear

It is all too easy to find fear-laden language about invasive species in the popular domain, but here I focus on an example from the halls of invasion biology itself. *Invasional meltdown* is a metaphor that can be traced to a paper in the flagship journal of invasion biology, *Biological Invasions,* by the ecologist Dan Simberloff of the University of Tennessee and his graduate student Betsy von Holle. Simberloff attributes the metaphor to a suggestion by an ecologist colleague, Peter Kareiva, and to familiarity with another metaphor in conservation biology, mutational meltdown, which describes how mutations may accumulate in small populations of a species at an increasing rate and lead to their extinction. An invasional meltdown, by contrast, concerns the effect of invasive species. The metaphor specifically refers to the "process by which non-indigenous species facilitate one another's invasion . . . potentially leading to an accelerating increase in number of introduced species and their impact."[4] Why was this metaphor chosen?

Having interviewed Simberloff, I suggest that he adopted this metaphor in part because he felt it was apt. Like Hebert and the barcode metaphor, he revealed in our interview that he has become a realist with regard to invasional meltdown: "It's a

pretty accurate metaphor. . . . I view it almost the way we picture real meltdowns, as this expanding mass of particles hitting
one another, and some of them are black particles and some of
them are white particles, and they are introduced or native, but
they're all interacting. That's just the way nature is." Not only is it
constitutive for him, but he has documented how it is "routinely
considered in various explorations by conservationist biologists, ecologists, and invasion biologists," indicating its broader
constitutiveness.[5] In contrast to Hebert, Simberloff claimed that
he "wasn't committed to people ending up believing this was
one of the major forces of invasion biology." Yet he did express
some desire to appeal to an audience: "I wanted people to read
it and think about it. I wasn't aiming at the popular domain. I
never thought there'd be anything in a newspaper about it at the
time. But I didn't want it to be just another paper . . . just sitting
there with no one bothering to read more than the abstract."

Simberloff also adopted the meltdown metaphor before it
had full empirical support. His retrospective, six years after he
coined the term, acknowledged, "A full 'invasional meltdown'
. . . has yet to be conclusively demonstrated." There are still only
a few well-demonstrated cases, including the interaction between yellow crazy ants and scale insects on Christmas Island.
Simberloff predicted in the interview, "I'd be surprised if we
didn't see a number of other cases within ten years," whereas
another ecologist, Jessica Gurevitch of Stony Brook University,
countered, "The lack of evidence for its existence is certainly
not for want of attention to the hypothesized phenomenon. . . .
[It] may, in fact, be uncommon." She continued, "It is alarmist,
but is it unrealistically so? We cannot know that until we answer the scientific questions of its generality and magnitude."[6]
Without having this evidence in hand, why would a scientist
prospectively adopt this metaphor?

First of all, Simberloff is an eminent ecologist, which allows him—as we saw with Hebert—to coin a novel metaphor. It has often been observed that such scientists, later in their careers, become more politically and socially involved. They have greater authority to sell their ideas and less to lose in terms of their careers. They can be more flexible about the usual restriction on values in science and can attempt to sell an idea with a metaphor that seizes attention. Although Simberloff claimed that "there is no evidence for an effect on credibility," this merely demonstrates the stock of credibility he has built up over decades, in contrast with young scientists.

But this still fails to explain why Simberloff would use this metaphor at all. I think his outspoken views against invasive species are the key here. Returning to his description of an invasional meltdown above, note that "real meltdowns" are metaphorical too. Also note that he invoked a black-and-white dichotomy between native and introduced species despite its philosophical problems.[7] Although there are neither good nor bad particles in an actual meltdown, he extends the metaphor here in a way which insinuates that introduced species are negative. These species are undoubtedly interacting, but does their interaction have to be a meltdown? In fact, the phenomenon under investigation is the development of new associations between invasive species, which could be seen as a harmonious process—it is certainly a normal biological one.

We should not be surprised that this term was used. Simberloff rejected an alternative, positive feedback, as "one of a number of boring titles" that lacked the negative everyday resonance of meltdown. Meltdown is a term that evokes concern and fear. It continues the pattern of using apocalyptic and fear-laden imagery and metaphors in environmental science, which includes Garrett Hardin's "tragedy of the commons" and Paul

Ehrlich's analogy between someone popping rivets on the wing of an airplane and the threat of a mass extinction. Simberloff acknowledged that it is "certainly pejorative": "'Meltdown' first appeared in 1965 with reference to nuclear reactors and, in the wake of the Three Mile Island disaster, became increasingly widely used, even metaphorically, to describe processes irreversibly deteriorating, apparently at an accelerating rate—children's temper tantrums, escalating internet crashes, the 2005 University of Tennessee football team, and the like." As Gurevitch described it, a meltdown "implies that after a certain point is reached, ordinary intercession is impossible, and a drastic state change is inevitable—whether it occurs in a toddler in the supermarket, a nuclear reactor, or an invaded ecological community."[8] These reflections by two scientists reveal its common associations, and it is to these that people will respond rather than to any purportedly technical meaning.

Even though the metaphor to a large extent preceded the availability of sufficient evidence, it has now been used for more than a decade, influencing people through the media and other outlets. Simberloff observed, "The meltdown metaphor attracted great attention, not only among invasion and conservation biologists, but also in the popular press." Yet it almost seems as though its underlying values in an extensive metaphoric web, rather than scientific evidence, justified its use. It is apocalyptic with regard to invasive species and thereby advocates for native ones. So, we effectively have a scientist adopting a metaphor to drum up concern about invasive species. What kind of influence is this metaphor likely to have?

The Discontents of Fear

The question of whether a negative metaphor such as this one is likely to inspire action toward invasive species is a complex, social scientific one. This complexity can be obscured by assuming that the problem has been properly framed as a biological one, so that all that matters is better scientific analysis. For example, Simberloff stated, "Response of any kind was unlikely without recognition of the problem in the first place, and to the extent that the meltdown metaphor has both fostered recognition and led to research elucidating the mechanism of the phenomenon in several instances, it has surely aided response." This claim assumes the linear model of science-society relations and flies in the face of extensive evidence that more scientific information and even better public understanding do not lead directly to desired outcomes, such as more action.[9]

In deciding to use this metaphor, Simberloff and his colleagues assumed that it would be effective. The focus was science, however. Although Simberloff remains convinced that it is epistemically useful, he admitted that he was unsure within the public domain. He claimed, "I don't have enough expertise in psychology and sociology," and "I'm not an expert in what the public understands," even though the decision to use the metaphor presumes such expertise. Rather than conducting empirical work to substantiate the potential effect of this metaphor, as a social scientist might approach it, we see folk psychology enacted. It is not a scientist's intuitions that matter so much, however, as the actual effects of a metaphor. As we saw above, meltdown has a decidedly negative resonance. Its emotive associations are fear-based. How will this influence people?

For the time being, until we have specific data on people's response to the meltdown metaphor, one place to begin is to assume that their response will conform with what we know about the effects of a "fear appeal." The literature provides ample reason to doubt the assumption that fear appeals will cause people to be more concerned about invasive species and—more important—to act on that concern. They may simply not work. They can be maladaptive if they lead to apathy and confusion rather than action. Social psychologists have shown that "extremely intense language or images used for purposes of persuasion can have an opposite effect upon the receiver," which they have christened the boomerang effect. Turning to invasive species, the social scientist Paul Gobster has reviewed previous research in persuasion and social marketing, mainly in the area of health, and shown that using fear-based appeals to create a sense of urgency to motivate action can be effective, but only if people are given something they can do. This is the crux of the matter, for otherwise effects on awareness and attitudes may not translate into action. Overall, he questions whether the negativity of invasion biology might cause the general public to think that "the main goal of the field is killing," so he has proposed that we substitute more positive, restoration-oriented metaphors.[10]

Other recent research, particularly in the pertinent area of climate change, supports the general conclusion that we should be wary of fear appeals. Alarmist language offers a thrilling spectacle of impending disaster, but it ultimately distances the public from the problem. Two social scientists, Saffron O'Neill and Sophie Nicholson-Cole, expanded on this theme in their review of fear appeals used for complex environmental issues, as opposed to more simple health-related ones on which most research has been conducted. They demonstrate that such appeals can desensitize, lessen trust, and evoke feelings that contribute to denial and apathy. Unless contextualized, fear-based communication disempowers people. They concluded, "Dramatic, sensational, fearful, shocking, and other climate change representations of a similar ilk can successfully capture people's attention to the issue of climate change and drive a general sense of the importance of the issue," but "they overwhelmingly have a 'negative' impact on active engagement with climate change."[11] To avoid these negative effects, they recommended that such communication be used quite cautiously; echoing Gobster, they advocated instead "nonthreatening imagery" to motivate genuine change. Although we don't know whether this climate change research is transferable to the subject of invasive species, it should at least give pause to some of the alarmism in invasion biology.

In addition to ineffectively motivating people, this language may also undermine scientists' credibility. Simberloff felt that "there is no evidence that this hyperbole has impeded scientific understanding or caused loss of scientific credibility." Yet such hyperbole can be seen as a form of stealth issue advocacy, which may affect the credibility of scientists and thus their capacity to contribute to environmental decision making. Pielke, for example, concluded, "It is when advocacy is couched in the purity of science that problems are created for both science and policy." The problem in this context is that scientists are presenting value-laden statements as scientific facts rather than engaging people in open dialogue about the dynamics and consequences of these species. Nelkin warned against such manipulative use of promotional metaphors, recognizing that the desire to speak broadly to the public even outstrips idealized scientific caution about pronouncements.[12]

To obtain further insight into people's response to such language in invasion biology, I conducted a few surveys. First, consider the results of an informal survey of students I taught in an introductory environmental studies course at the University of Waterloo. After the course, 90 percent of them were concerned about the environmental impact of invasive species, yet over half felt that it was inappropriate for scientists to use the term invasional meltdown, given that it has not been "conclusively demonstrated."[13] Such data, though preliminary, lend credence to concerns about how the meltdown metaphor might affect credibility in the public domain.

Second, a student and I conducted an online survey of invasion biologists themselves that found similar concerns from their side. Although 27 percent of the 420 respondents agreed that "hyperbolic language about invasive species is needed to capture the attention of policy makers and the general public," nearly twice as many disagreed with this statement. There isn't a consensus here, but many of them clearly imagine better ways to engage the public on issues of invasive species. They are definitely concerned about hyperbole, 61 percent agreeing, and only one-third as many disagreeing, with the statement "By using loaded language, invasion biologists may erode public trust in their objectivity."[14] In contrast to Simberloff's belief, we see here some trepidation about using loaded language. Nonetheless, it is possible that such trepidation reveals a relatively conservative view of the role of science in society.

A more sober question is whether it is deceptive and manipulative to induce fear, perhaps in a manner we might not expect of the scientific community. Simberloff claims that meltdown draws attention to the issue, yet the metaphor needs to be interrogated because it communicates more than just the facts of the matter. Some scientists may interpret it as a technical issue that isn't subject to public values inquiry, even though people are being swayed by the emotive overtones and suggestiveness of the language. This is not to say that lay people will be duped by discourse. Nonetheless, such a metaphor is morally questionable because it increases general anxiety among members of the public that they cannot resolve, as they have very little capacity to do anything about it. A paper in *Psychology and Marketing* argued, first, from deontological ethics, that this is always inappropriate, regardless of any social consequences of not using fear, and, second, from utilitarian ethics, that it is appropriate only "if it produces a net balance of good over bad . . . *and* if other approaches are less effective." In the case at hand, as far as I can tell, these questions were not addressed. A related issue is the effect of fear appeals on children, who might be particularly vulnerable to having their disconnection from nature and incipient fears of it being reinforced by such messages. Children may become fearful of the natural world quite easily, and such fear leads them to dislike natural areas.[15] This is a risk we cannot afford to take with the next generation.

By manipulating emotions, the meltdown metaphor also maintains the social authority of those who seek to treat the problem. The situation is analogous to a hierarchy that developed in the medical realm. As the Australian environmental sociologist Catherine Allan explained, "Illness is an emotionally charged subject, and the fear, anxiety, guilt, hope, and denial associated with human illness can be transferred to discussions about natural resource management, leaving people vulnerable to persuasion and manipulation, especially from the experts who have the training, knowledge, and power to treat the perceived illness."[16] Although metaphors may create the illusion that there is no power hierarchy between scientists and society, they may in fact strengthen this hierarchy by blurring the boundary between persuasion and the findings of scientific research.

Rather than this authoritarian approach to communication, we might adopt one based on dialogue. In the introduction to their edited volume on how we might more effectively communicate to inspire action against climate change, Susanne Moser and Lisa Dilling defined communication that supports social change as "a continuous and dynamic process unfolding among people that facilitates an exchange of ideas, feelings, and information as well as the forming of mutual understanding and common visions of a desirable future."[17] Instead, scientific use of a fear appeal institutes a hierarchy between scientists and the public. It is assumed, as it has been in the war on terror, that the right decision has been made and that those in power simply need to act, even if substantiating evidence is limited.

Another indication of this social hierarchy is that scientists blame misunderstanding on the press and the public, rather than on their own choice of metaphor. Simberloff claimed that he "thought about all the right things," yet he did not see his metaphor as either emotional or controversial. He also acknowledged in the interview that he "didn't think about" whether he might have set himself up for misinterpretation by using the metaphor he did. He would have used the metaphor anyway, because "the press always misrepresents science." As he stated elsewhere, "Some [writers for the lay public] have stretched it well beyond its meaning as understood by invasion biologists."[18] By adopting this view of science communication, scientists disclaim responsibility for their linguistic choices, even though the interpretations might well have been predictable. The conduit model is at work. The net is being raised and lowered to accord with a preferred definition: some associations of meltdown are considered scientific, whereas others are not.

It's even possible, in this context, that scientists purposefully adopt metaphors that are likely to be misinterpreted. Simberloff reported, "It is true that martial metaphors occur in the invasion biology literature [but] such metaphors become more vivid and pervasive when the popular press reports on these subjects,"[19] ignoring how he let this particular cat out of the bag. Metaphoric ambiguity allows a scientist to maintain the authority to create and use language to inform, while at the same time remaining out of the resultant political fray. It is a way to avoid political debate about your values—just claim that you are presenting the objective facts as they are, as if they don't have to be debated, when in fact your position is replete with value-laden implications. Simberloff's own words in his recent review paper instantiated this metaphorical paradox when he referred to meltdown as a constitutive metaphor at one point, emphasizing its scientific merit, and to hyperbole at another, emphasizing its subjective nuance. It is a metaphor capable of serving diverse roles, but the pertinent question is whether a fear appeal will engage people with the issue.

Questioning Militarism

As we know from 9/11, one of the best ways to incite war is through fear. Fear of invasive species and their effects—to which the use of the word meltdown contributes—leads people to oppose them—often militaristically. Militaristic metaphors are rampant in popular reporting on invasive species, but they also occur in the scientific literature, where major hypotheses for their spread include the "enemy release" and "novel weapons" hypotheses.[20] Such metaphors are performative, encouraging weed-pulls and control programs, the erection of barrier zones, lucrative contracts for herbicide companies, and research grants for invasion biologists. It is an oppositional mindset in which we defend a certain view of nature, one from which humans have been exiled.

Invasion and militarism are feedback metaphors linked together by a narrative driven by fear. Like fear, however, this militarism is not without its socioecological drawbacks. But before touching on these, I first acknowledge the presumed benefit of militaristic metaphors: they may encourage conservation action and attract research funding to study invasive species. The psychologist Robert Ornstein and the ecologist Paul Ehrlich proposed that our species' adaptation to short-term as opposed to long-term thinking is nonadaptive in the face of our contemporary influences on the planet.[21] We have evolved fight-or-flight reactions to predators that appear suddenly, but we cannot see the survival implications of less evident issues, such as our effects on the environment. Militaristic metaphors attempt to activate a response to get us to do something immediately.

Militaristic language is commonly used in environmentalism, a strategy that dates back to before Rachel Carson's *Silent Spring,* in which it served as a clarion call that helped initiate the environmental movement.[22] Ehrlich's own influential metaphor, the population bomb, exemplifies this strategy. In the case of invasive species, we may require strong language to draw attention to the issue, because it is otherwise unapparent to most people. Certainly, there are times when this language and approach are appropriate; past use alone, however, is insufficient reason for using them uncritically today.

Numerous scholars, including invasion biologists, have begun to question a militaristic way of relating to these species. One popular book on invasive species, for example, sought to use common words rather than be tempted into "linguistic maelstroms of war and pestilence."[23] Some people might claim that questioning militarism is mere apathy, but this dismissal simply normalizes the perception that you are either with them in this war or against them. Rather, it is one way to attend to the industry built up around how we conceive of invasive species. For example, an editor rejected my first paper on this issue, without review, commenting, "The bottom line for me is that, given the abundant, massive, and seemingly insurmountable global conservation problems that we face, the semantics of dealing with invasive species is a low priority."[24] Not only does this misunderstand semantics and the performativity of language, but it fails to grasp the interdependence of science and society. When editors gatekeep to this extent, the underbelly of an academic worldview is being protected.

Let's examine some reasons for questioning a militaristic emphasis in invasion biology.[25] Some of them, introduced earlier, relate to the appropriateness of inducing fear. For reasons related to the boomerang effect, militarism may not even be effective. The overuse of war metaphors may lead to their becoming vapid when we really need them, like crying wolf. More broadly, although these metaphors may effectively motivate conservation action in the short term, they could be ineffective in a long-term socioecological context. To assess this possibility, we need to consider how we have framed the problem, attending to the boundaries we have placed around it and to the observers. Ecology lends itself to such systems thinking, yet there remain ways that invasion biology retains antiquated dualities. When the models we have are reductionist and control-oriented, they may be inadequate for dealing with complex, emergent systems. To deal with the latter, we need to change how we think.

A militaristic approach to invasive species relies on a duality between nature-nonhuman-native and culture-human-nonnative that excludes human beings from ecological systems. There is nature-without-humans and nature-with-humans. We take invasive species as an indicator of the latter, and thus we fight against them. We may not want them somewhere, whether it is because they are taking over our yard or overwhelming a nature reserve, but that is tied to a particular way of perceiving the situation rather than being an immutable insight. Many philosophers and social theorists would turn the underlying duality on its head by demonstrating that we continue to hold on to this duality even though hybrids between nature and culture surround us on all sides. The Finnish ecologist and environmental policy scholar Yrjö Haila described the general problem: "We want to know what nature allows us. To reach an ultimate certainty we would like to distinguish between 'nature by herself as a standard' and 'nature modified and polluted by humans.' This, however, cannot be done. Humans are creatures of nature; consequently, discriminating between phenomena of nature as 'natural' and phenomena of culture as 'unnatural' does not make sense at all."[26] By framing invasive species as a problem out there, however, we are misled into overlooking how they are a symptom of ourselves. We are embedded within invasive species, from their causation to their formulation as a problem. Yet we forget that we have created them and that they cannot exist if we are not active as participants who move them around and as observers who value in a particular way. Through these forgettings, invasive species become targeted as the enemy, and battling them becomes the solution.

Like it or not, we live in a hybrid world where even "native" landscapes have been shaped by humans. They have been shaped by direct disturbance both from humans and from our livestock, or from other effects we have had, including acid rain, exotic earthworms, global warming, nitrogen deposition, and reduced fire frequency. These disturbances contribute to and even encourage the spread of invasive species. We are thus

mistaken to treat invasive species as the primary cause of ecological changes we are seeing. I think we do so because they are
sometimes easier to deal with than overarching causal factors.
A number of recent ecological studies have also shown that
exotic and native species are more similar than we assume,
having classified them as opposing armies in a battle. Invasive
species may even play positive ecological roles, as they have in
some restoration programs.[27] Invasive species are now part of
the way the world is; they are not independent of us. Our habits
and our disturbances are often the cause of their introduction
and subsequent spread. Consequently, the duality represented
by a fervent war may be too simplistic a model for our approach
to these species.

A short vignette may help demonstrate the conceptual
problems involved in protecting native ecosystems against the
threat of invasive species. Consider the invasion of emerald
ash borer in southern Ontario over the past decade (Figure 14).
Undoubtedly, its introduction has had aesthetic, ecological, and
economic repercussions. Can we really make a case, however,
that it presents "an enormous threat to *native* ash forests in eastern North America?" In so doing, we simply demonstrate the
flexibility of the term *native* and how we adapt its meaning according to context and rhetorical needs. If the borer is invading
a native ecosystem, then how do we differentiate the forests of
fifty years ago that had American elm as a dominant species, or
the earlier forests typified by American chestnut? Both of these
trees largely succumbed to invasive pathogens. Furthermore,
ash was overabundant in these heavily disturbed, early succes-
sional landscapes, which provided conditions supportive of the
beetle's outbreak. We may regret the loss of a primeval forest,
and we may bemoan continued losses, but regardless, we cannot say that we are losing something timeless or native. It may

Figure 14. A woodlot in southern Ontario, Canada, in which
the ash trees that formed the canopy have all been killed
by emerald ash borer.

be helpful to differentiate a continuum of human disturbance
and naturalness, however we define them, yet there is no self-
evident point at which it is sensible to say that an ecosystem is
native.[28] This classification must occur as part of a larger social
discussion because it is not an issue restricted to the realm of
ecological science.

The possibility of such discussion, however, becomes
mired in interpersonal and social conflict created by the aforementioned duality and resultant militarism. The language of
war derives from a strong moral commitment that tends to po-
larize not only villain and victim but also those who oppose the

war and those who support it.[29] Rather than being encouraged to see shades of gray, then, we are left with a black-and-white opposition between nature-with and nature-without invasive species. Those who battle them are on the right side, which validates a war at any cost, including interpersonal cost. Militaristic and other forms of polarizing metaphors and imagery thus contribute to environmental conflict.

In October 1997, for example, the California Department of Fish and Game announced its plan to apply pesticides to a water reservoir to exterminate invasive pike. Because the reservoir supplied fish and water for the nearby town of Portola, local people were outraged that this was a war—not against the fish but against them. In another example, consider the National Park Service project to eradicate rats from Anacapa Island in the California Channel Islands. Although the decision to act was undoubtedly a difficult one, a local animal rights group interpreted the method chosen to control the rats—precision dropping of poison pellets along GPS gridlines by helicopters— as a symbolic and technological act of war. They thus opposed the eradication project despite some shared objectives with the Park Service; cooperation might well have been possible. And in a final example, soon after the emerald ash beetle was detected in southern Ontario, a government department, the Canadian Food Inspection Agency, oversaw the cutting of a ten-kilometer-wide swath of healthy ash trees to create a "barrier zone" to prevent its spread. Most of these trees were cut on private land and with little notice to local landowners, which contributed to tremendous conflict, including physical threats.[30] In each of these instances, the rhetoric of bioinvasion contributed to human conflict because the agencies involved were so committed to their actions against invasive species that these actions became a perceived affront to a particular group of people.

Militaristic language obscures the values and associated views of nature held by those of us observing invasive species. People weigh different values in reaching decisions about them. I am not a deconstructivist who denies the existence of nature "out there," but at the same time I would emphasize that how we relate to this nature depends very much on our particular cultural context. Invasive species do not necessarily have a negative effect on cultural practices, for example; in fact, some of them may "augment cultural traditions, through their inclusion in lexicons, narratives, foods, pharmacopoeias and other tangible and intangible ends." As just one example, rural peoples in the Eastern Cape, South Africa, would prefer higher densities of certain invasive species, both for consumption and as a source of income. It follows that in the long run we will need to better understand such stakeholder values before managing invasive species.[31] Even if we conclude that they must be exterminated in a particular locale—and quickly—consultation will lead to better outcomes. Thoughtless adoption of militaristic language and planning could instead spark conflict between those on the two sides of such an issue.

Consider next the common charge that policies against invasive species are xenophobic. Such an argument may appear tortuous, given that these species can have real effects, yet we must keep in mind that anti-immigration policy just as readily cites the "real" problems created by immigrants. Although biologists can rattle off numerous reasons that invasive species are not like immigrants, a larger metaphoric web influences how discourse about them is interpreted. I would be the last to accuse conservation and invasion biologists of being racists because of their concern about invasive species. Rather, we have to acknowledge the extent to which this issue interacts with the history of racism and xenophobia in the United States.

For example, the U.S. media commonly invoke metaphors that interpret immigrants as animals, so we might expect that the reverse metaphoric mapping, interpreting nonnative animals (and plants) as immigrants, can be interpreted as xenophobic. Echoing the thesis of this book, one observer concluded, "The commonly perceived separation of natural and human spheres that allows us to view derisive antiforeign rhetoric as 'benign' when applied to nature, but as 'malignant' when applied to humans, cannot be sustained because the images they conjure and the meanings they carry are inseparably entangled, and thus one can never be certain what sentiments lie behind an expression of concern about 'the exotic.'"[32] A war against exotic invaders cannot be isolated within the realm of invasion biology.

At a broader and more speculative level, war is war in the context of the metaphoric web. When we use militaristic metaphors, we create a similarity that contributes to a semantic field of war. In the words of Daniel Nelson, a senior fellow at the Center for Arms Control and Non-proliferation in Washington, "Human conflict begins and ends via talk and text." It is thus disconcerting that there has been a recent switch by the media to presenting real wars with nonmilitaristic metaphors, while transforming daily occurrences into wars with militaristic metaphors. Through recent wars against AIDS, cancer, drugs, poverty, and terror, we diminish the significance of actual war. A war against invasive species thus supports veritable war, indirectly, by implicitly endorsing and reinforcing the logic of war. In the mid-twentieth century, for example, killing insects with insecticides became interwoven with killing humans in war. Given the tremendous socioecological effect of wars around the globe, we need to be cautious with language that contributes to polarized, militaristic ways of conceptualizing difficult situations. Instead, we need to reflect on alternatives that are more consistent with a vision of sustainable socioecological systems. As Cheryll Glotfelty, of the Literature and Environment program at the University of Nevada, Reno, noted: "Perhaps in Carson's day, war was a necessary and appropriate context in which to conceptualize environmental issues. But, thankfully, the Cold War is over. Should people who are committed to enlightened stewardship of the earth continue to invoke it?"[33] We need to consider ways of reframing the issue that invite other perspectives into the conversation, and which don't mislead with distorted and exaggerated language.

Alternative Perspectives

Using fear-inducing and militaristic metaphors, invasion biologists advocate for an oppositional response to invasive species. Although this is sometimes appropriate, I suggest that in the long run we require substitutes that ripen our relationship with each other vis-à-vis invasive species. As Pielke has recommended, invasion biologists need to be honest brokers of policy alternatives, presenting the costs and benefits of different courses of action using different metaphors. Not only would this maintain their credibility, but it would also allow them to expand policy options rather than closing them down, as advocacy tends to do. We know from cognitive psychology and conflict management that we tend to rely on preexisting cognitive maps to reduce the effort we require to think things through, yet this narrows the options for solutions too quickly. Instead, it has been shown that there are benefits to maintaining the options a little longer to experiment with different ways of viewing a problem.[34] We are still in the early stages of grappling with invasive species, so I propose that this is precisely what we need to be doing.

As an exercise in problem reframing, I've been experimenting with alternative metaphors that highlight different aspects of this phenomenon. At first, it may appear that they are difficult to find. This merely demonstrates the extent to which our current view is entrenched. Further, the purpose of seeking novel metaphors is neither to reject entirely the ones we have nor to provide a single alternative.[35] Instead, new metaphors reveal limitations with the usual view of invasive species and thereby point to elements of more creative thinking about them. They enrich our perception. They can break the grip of ingrained opposition so that we have a full range of options and better understand the trade-offs involved in acting versus refraining from acting. If we can only oppose these species and the ecological changes they bring, we will be stymied by the changes that are nonetheless happening and unable to accept them. In the long run, we require expansive and socially embedded thought about these species because they will not go away. Some are here to stay, no matter what we do. And other new species will continue to arrive, and we need to have better ways to decide what to do about them in a social context.

You may have noticed that I sometimes refer to these species somewhat vaguely by not specifying that they are invasive. I do this so that the notion of invasion is less active. Analogously, Lakoff recommended that Democratic politicians in the United States refrain from referring to the Clear Skies Act because its very name would insinuate that it was a progressive amendment to the Clean Air Act that would reduce air pollution—when its probable effect was quite the opposite.[36] Similarly, by referring to invasive species, we reinforce the very frame that is in question here. We probably won't extinguish this language anytime soon, yet we can still begin to shift our perspective with the minor intervention and experiment of discontinuing to use it. Otherwise, we activate the negative connotations of invaders, and the only alternative to doing something seems to be its polar opposite, befriending them unconditionally, which is equally unsatisfactory.

A related alternative is to adopt more neutral, objective language. It is just this approach that has been promoted by some invasion biologists who feel that the language of invasion biology has become too militaristic or too subjective. In a recent paper in *Biological Invasions*, for example, the invasion biologists Robert Colautti and David Richardson claimed, "The indiscriminate adoption of societal values into scientific terms and definitions in primary research articles . . . is less efficient for the advancement of scientific theory." This proposal for a neutral language may seem consistent with what I'm proposing; it might, for example, serve to maintain scientific credibility. Yet novel thinking about these species will not come from viewing scientific inquiry as if it occurs in a social vacuum.[37] We must negotiate the Scylla of apathy and the Charybdis of advocacy to find more creative approaches.

These invasion biologists justify such neutrality with the assumption that their science will progress more rapidly if it avoids unclear and vague language. In contrast, a recent review concludes that science may progress despite vague language for a variety of reasons, especially in an immature science such as invasion biology, where we may not understand a phenomenon well enough to be more precise.[38] This of course does not deny that we need a certain level of clarity for communication and investigation. The more problematic assumption, as we've seen before, is that such clarification will lead directly to better social outcomes. It is our old friend the linear model, which intimates that the contribution of scientists to policy is to provide the facts of the matter. In the current context, this approach can be seen as a different way to camouflage the linkage between facts and values: pretend there are no values and just talk facts. It is stealth issue advocacy of a different sort, through neutrality rather than resonance. In this way, it is again top-down and reflects scientists' values and priorities rather than their actually engaging with society. The experts speak and thus we can dispense with democracy.

We need to question this epistemic rather than social emphasis, including the recurring assumption that these dimensions are so easily parsed. Colautti and Richardson, for example, reiterated the fact-value dichotomy by introducing an "artificial dichotomy" between two neologisms, *motivational subjectivity* and *methodological subjectivity*. The former is the realm of science and conceptual questions and limits subjectivity through the constraints imposed by ecological theory. In contrast, the latter has more to do with the way terminology motivates scientists or inspires the public and helps them understand. The authors urged, "Invasion biologists can best serve society if [they] allow motivational subjectivity into public discourse but attempt to minimize it in the methods of scientific inquiry." I doubt anyone would argue with this intention, but this book serves as an argument for why it is a pipe dream, given the language environmental scientists actually tend to use. To sound the death knell of such positivism, the respected Princeton philosopher of science Bas van Fraassen once stated, "The language of science cannot be reduced through 'operational definitions' or translation into a hygienic, pure observation language. . . . Demythologizing the language of science is impossible." Although there is more and less neutral terminology, such issues of definition, whether meant to be neutral or not, are undeniably political.[39] It is also noteworthy that such pronouncements are given—as we've seen in the case of other scientists—with no reference to any supportive evidence. The two neologisms mentioned above inscribe the fact-value dichotomy without engaging with any of the pertinent philosophical literature. We will not obtain optimal social relations with invasive species if biologists continue to present themselves as experts on issues of communication and philosophy.

If scientists want to have relevance to our interaction with the natural world, they need to walk the interface between subjectivity and objectivity with greater nuance rather than merely reacting against one extreme by returning to the other. At least Simberloff's *invasional meltdown* engages with society, demonstrating in itself the illusoriness of the motivational-methodological divide. Motivation is always there, even underlying the conduct of any method oriented toward comprehending invasive species, though sometimes unacknowledged. As two ecologists, Gay Bradshaw and Mark Bekoff, have observed, "The subjectivist states his judgements whereas the objectivist sweeps them under the carpet by calling assumptions knowledge." Invasion biologists belong to a crisis discipline, as conservation biology has been called, because they have concerns about biodiversity.[40] This must be acknowledged. Otherwise, their field is cynically a branch of ecology posturing as something different to appeal to funding agencies. In this context, scientific disengagement through neutrality would be disingenuous because it maintains the ideal of having it both ways: objective language to allow invasion biologists to separate themselves from society and allow the progress of their field (narrowly defined, rather than socioecologically), while at the same time permitting them to obtain continued funding for their work, which requires that it have at least some relevance to management concerns. This approach maintains their authority on the topic, even while fencing it off so that it only indirectly relates to real societal needs and interests. It is for all these reasons that we must reject the notion of a neutral language for invasion biology or any environmental science.

We can now ponder more explicit alternatives to fear appeals, militarism, and our usual way of conceiving these species. Many have been proposed, yet as I will discuss in the next chapter, my role is not to adjudicate among them. Instead, I wish to shake up our usual perspective. For example, I have reviewed thirteen divergent ways we can conceive of this phenomenon to highlight its different aspects. Nearly every time I speak with a scientist who has thought about these species, though, I encounter new metaphors. They have been reconfigured within the framework of human security, or by analogy with the human immune system, or as passengers rather than drivers of environmental change. They have been thought about with metaphors such as pressure and resistance, ecosystem saturation, and ecosystem health. In toto, this conceptual search suggests that others recognize problems in the paradigm of invasion biology. As the distinguished geneticist John Avise of the University of California at Irvine acknowledged, in his search for more apt genomic metaphors, "Metaphors can and should evolve to accommodate new findings."[41] As we learn more about these species and their social context, our metaphors must evolve. The discussion of human security, for example, explicitly attempts to overcome the inaccessibility of some ecological language and replace it with a framework that allows diverse stakeholders to contribute to the discussion.

As I stated earlier, new metaphors will make only partial inroads if they still adopt the paradigm of invasive species. In our search for alternatives, we need to attempt to reconcile some of the shortcomings cited above, in particular those metaphors that engender a duality between people and the natural world,

Figure 15. This sign along the Trans-Canada Highway
demonstrates the challenge of preventing invasive species from
dispersing unless we are willing to modify our habits dramatically,
including our reliance on global networks of commerce.

one reflected in interpersonal conflict. We need to think more
of systems and socioecological sustainability. We also require
metaphors that recognize that ecological systems are in flux and
that we cannot control them entirely.

Our conceptualization of invasive species relies on a duality between inside and outside. We can see this in how the
current scientific literature emphasizes the propagule pressure
exerted by invasive species and the opposing biotic resistance

of native communities. The inside here represents a concentration of native species that we want to protect. Instead, we might reorient toward modified systems' ability to maintain ecological functions that are important to us. These two perspectives represent alternative normative concepts in conservation biology; the former, compositionalism, emphasizes biodiversity and thus tends to exclude humans from nature more than the latter, functionalism, which instead focuses on concepts such as ecosystem health and sustainability. Ultimately, we require elements of both approaches because the former is better suited to reserve areas and the latter to more urbanized and disturbed ones. Although disturbed areas may not maintain endemic species, they may nonetheless fulfill functions such as carbon sequestration quite adequately.[42] Our metaphors need to keep such issues on the table.

We could also try slight definitional shifts. Invasive species are ones that become too abundant, just as some native species do. We might recognize this connection by referring, in both cases, to "superabundant" species. In other contexts, we might focus on the fact that our specific concern with superabundance is its effect on human preferences. These species are in fact "harmful" ones, a term that would encourage more open discussion about what we expect of nature, in particular through more explicit recognition that it is our interests that are often at stake.

Consider a more specific example. It has been proposed that native species' normal behavioral patterns are maladaptive in the presence of new, nonnative ones, an "evolutionary trap."[43] In cases where the latter cannot be eliminated, it may be possible to "inoculate" naive native populations with individuals from populations that have been exposed to the new species to allow them time to adjust evolutionarily and thus avoid extirpation. Though controversial, such a view seeks long-term coexistence with new species rather than an unrealistic return to a pristine former state without them.

More radically, other conceptions might better highlight the connection between the spread of these species and human globalization and cosmopolitanism. When we think about invasive species, we typically focus on Band-Aids and other stopgap measures, rather than potential system faults related to human behavior (Figure 15). An analogous situation exists in our approach to climate change. In particular, our focus on carbon production ignores a much greater issue, our consumption. We need in both cases to focus instead on the roots of the situation. This requires that we redirect our focus from the problem out there to the much more challenging one discussed here, that of the relation between different people holding differing views of nature.

With courage, we might recognize that our fears about these species will lessen over time, which may help inspire very different approaches toward them. We can see such a process in the gradual acceptance of species that have been in particular places for long periods. It has also been observed in relation to climate change. Mike Hulme, a professor of climate change at the University of East Anglia, for example, reviewed the different discourses of fear about climate change through the ages, concluding with the current apocalyptic narratives. In each earlier case, he argued, fears were conquered over time through cultural change; for example, fears that climatic change were God's judgment in the Middle Ages were tamed by naturalistic explanations of weather. He observed that our efforts to conquer climate change now tend to rely on mastering climate through various forms of control and engineering, whether technical or social, which "suggest that climate is an objective reality to be manipulated through material intervention. They imply an unambiguous separation between Nature and culture." Instead, he proposed that it is "only through further cultural change, working on and through material processes, that the contemporary discourse of climatic catastrophe will be dissolved." He concluded, "Rather than seeking to conquer climate, we should be aiming to celebrate climate and respect it as part of ourselves."[44] We are a long way from celebrating invasive species, and there are some we will never celebrate, but our metaphors for them should help us recognize more deeply that nature includes us and that we are not totally in control.

VII
Seeking Sustainable Metaphors

For what is freedom, if not the ability to choose
by which metaphors we will be seized?
—*David Edge,* Technological Metaphor and Social Control

We can say that ethics, having once been a secret measure
of scientific truth, can now become its overt judge.
—*Paul Feyerabend,* Conquest of Abundance

I n previous chapters we have seen some of the challenges presented by feedback metaphors, but we have only touched on what we might do about them. We might wish to cultivate normative language, but how shall we do that? Which method shall we adopt? And which values shall we emphasize? The point is not to find a perfect metaphor, which does not exist, but to demonstrate that when scientists use a metaphor, they are endorsing particular values. I have contested the values associated with the metaphors of DNA barcoding and invasional meltdown, yet the use of such value-laden metaphors might nonetheless be closer to the type of science we want. It is conservation science that is more confident of itself and its message. A remaining challenge, however, is to include more people in conversations about the values we wish to promote with environmental metaphors—and thus about the type of future we envision.

In what follows, I shall not prescribe which values to adopt. I have presented what I consider to be key dimensions of where we need to be looking, specifically how kindling connections both among people and between people and the natural world can contribute to sustainability objectives. These dimensions are proffered as general guidelines; I believe they are reasonable ones that we can accept as a starting point. Many have advocated sustainability as a direction in which humans should aim, despite its flaws. And others have reviewed the limitations of dualistic thinking for approaching sustainability. Metaphors have much to do with connection, so they have tremendous potential for annealing dualities in the interest of socioecological sustainability. I nonetheless recognize that there are diverse, local priorities other than this one.

I hope to avoid the criticism of presenting solipsistic value preferences, which is a general challenge for critical discourse analysis. Undoubtedly, I am promoting a particular conservation-oriented and liberal view, one that is akin to the idealized egalitarianism noted among critics of metaphors in genetics.[1] As we will see, this may create its own challenges if the view presented here does not mesh with how most people view the environment. Thus, rather than proselytizing for a particular stance, my aim is to increase awareness and thereby initiate further discussion. This objective simultaneously avoids a second pitfall of promoting my personal preferences: presenting myself as the authority figure to replace scientists in choosing which values to adopt. Instead, I believe that such choices need to occur as part of a broader public discussion. As we open the lines of communication for these discussions, we will increasingly foster the human interactions that are so essential to sustainability.

None of this should be construed as political correctness in the disparaging sense it is sometimes understood. I hope to have highlighted large-scale, performative feedback metaphors, ones that truly have an effect. You may conclude that I am incorrect about their value implications, but the overarching message nonetheless holds: if we want to communicate effectively, we need to do so as inclusively as possible. We need to communicate our values clearly rather than hiding them within our metaphors. If we speak in a way that includes just our group, then we risk excluding people that may share our deeper values. Even if radical language change doesn't happen overnight, I propose that more gradual shifts in our language can also be effective in meeting our broader objectives.

We all have roles to play in this vision. Our roles will differ with respect to choices about introducing novel metaphors as opposed to continuing to use ones that are already established. The moment when a new metaphor is introduced should not be taken too lightly, as this is the time to act—before it becomes entrenched. I thus begin with a look at how people might participate at this early stage in the creation of environmental metaphors, rather than leaving these choices to scientists alone; we will all have to deal with their implications later on. I then consider the extent to which social scientists might be able to assist with such decisions about metaphors.

These two topics apply mostly to new metaphors, but they represent elements of broader metaphorical reflection and reflexivity. Such reflections should give rise to a sense of responsibility among both scientists and nonscientists, whether they are contemplating new or old metaphors. Though we cannot entirely control metaphors and their resonance, that is no excuse to use them heedlessly. Instead, the scientific community needs to relinquish the conduit model of communication and become more responsible for potential interpretations of its metaphoric choices. Yet this will have limited effectiveness unless nonscientists learn to reflect more deeply on their interpretation of scientific metaphors. Although this may seem idealistic, to me it simply represents a vision of science and society becoming more engaged with one another in the interest of sustainability.

Early Participation in Metaphoric Choices

Environmental scientists introduce new metaphors with some frequency. I cannot estimate this frequency with any confidence, though it is certainly a topic worth exploring. Regardless, my case studies show that scientists tend to adopt metaphors that resonate in their social context, and these metaphors can have significant implications when recirculated back into society. In particular, environmental metaphors are replete with values, values that need to be made explicit so that they can be discussed. These discussions concern us all.

Discussions about metaphoric choices should not be restricted to the experts, but opened to a broader array of people representing relevant interests. Numerous scholars have argued for a postnormal science that requires greater public involvement, given the increasing uncertainty, complexity, and fact-value linkage of contemporary environmental issues. This proposal does not deny that scientists can be experts on aspects of these issues. But it is increasingly acknowledged that change will come about only by recognizing the forms of expertise that nonscientists possess, thereby distributing expertise more broadly through society.[2] It is a vision of science working with society rather than just providing abstract facts for potential social uptake.

If we reconsider our implicit notions of expertise, we might be more motivated to increase public involvement in such decisions. Expertise is not limited just to scientists; there are instead different types of expertise. For example, there are researchers in a field who do "core" science, as opposed to those conducting research in different fields, as well as the public, and the middle group does not necessarily have any more expertise on a particular issue than the last one.[3] In fact, we might follow some scholars and recognize three pertinent types of expertise: no expertise (which we will set aside), contributory expertise, and interactive expertise. When you have contributory expertise, you are able to contribute to a discussion through knowledge that is either local or academic and scientific. In contrast, those who have interactive expertise demonstrate the ability to interact with those who possess knowledge in different areas. Scientists are no more likely to have this capacity than nonscientists, but nonetheless it is a critical one in discussions of environmental affairs that incorporate facts and values. Metaphors are a case in point: decisions about them require people with different forms of contributory expertise, as well as those who can facilitate discussion of the interweaving of facts and values in metaphors proposed by different individuals and groups.

Clearly, this is a different model from the linear model, where we await scientific facts to set us free. In contrast, Daniel Sarewitz has demonstrated that science can in fact make environmental controversies worse.[4] We approach them with multiple expert disciplines, each of them bringing its own perspective and values and leading to selective interpretation of complex multidisciplinary sets of facts. He thus maintains that we have an "excess of objectivity" that gives rise to "contradictory certainties." Rather than being antiscience, this view displays its actual richness. The resultant uncertainty might seem to necessitate more science, yet science cannot resolve this quandary. Instead, Sarewitz proposed that we bring the politics of science to the fore, rather than camouflaging it behind a particular interpretation of the facts. By revealing rather than camouflaging the underlying values, we can do the politics up front and then use science toward preferred endpoints. This model, a relative of the honest broker, contrasts with the usual method, by which we attempt to utilize science within a policy context in which it can be interpreted (if it is understood or adopted at all) in numerous ways.

Since scholars have recognized the need for greater public involvement in environmental decision making, there have been explorations of a variety of methods, including citizens' juries, roundtables, and other deliberative and participatory methods. Such processes provide an opportunity for open dialogue and debate among those having different perspectives in the interest of obtaining the best possible choice under the circumstances.[5] This choice arises from the interaction between diverse stakeholders who help construct knowledge and set priorities. Otherwise, value commitments are smuggled into scientific conclusions to the extent that facts are made to seem inviolate.

Without reviewing deliberative methods in detail, I wish to consider a few elements relevant to deliberation about environmental metaphors. This linguistic strand exemplifies the evolution of new forms of deliberation, and I propose that explicit attention to the moral quality of feedback metaphors such as those studied herein is a cutting edge for contemporary science studies. Numerous scholars have considered how we might make science more democratic and inclusive of values, but I wish to extend this engagement into a different, linguistic terrain.[6]

Novel metaphors will continue to be developed and introduced. I propose that deliberation needs to occur relatively early in the process of considering a metaphor, before its effects become clear. Furthermore, people will consider this exercise more legitimate if they are included early on, when decisions are made, rather than as an afterthought. For example, we may be able to develop novel participatory infrastructures to use digital media to increase stakeholder involvement in seeking alternative frames.[7] This intent aligns with the growing trend toward involvement of nonscientists in the early stages of decision making, which recognizes that this not only leads to better science but to science that is more palatable to diverse publics.

Some scientists might resist such a move because their considerations of the democracy of science reach mainly to issues external to the actual science itself, such as who gets funding and why. But too often this ignores how similar questions about the democracy of science can be applied to its very core.[8]

That is, we can question assumptions that are made about the democracy of scientific inquiry and methodology themselves, and, more important, question how they might be reformed. Metaphors, as we've seen, crisscross the boundary between considerations that might otherwise seem internal or external to science, so they are an area ripe for investigation. They may seem to be less risky than transgenic organisms or new nuclear technologies, but they can have dramatic real-world effects by laying the groundwork for human interactions over long periods. We require public involvement both in the technical realm to assess whether we can accept uncertainties associated with a new technology, and with the metaphors that lay the seeds for such implications. Thus, rather than dropping a new metaphor into society as they see fit and watching its widening circles afterward, scientists would engage with nonscientists earlier to obtain a better sense of public response. As we've seen, scientists' intuitions are not necessarily our best guide.

How will we decide who contributes to decisions about metaphors? It would clearly be unwise to rely on a simple majority vote, if that were even possible, because there are numerous limitations with such "vulgar democracy."[9] Individuals may lack necessary understanding or the commitment to become engaged in a meaningful way, though this is not a statement in favor of a naive version of the deficit model. Just as we cannot take the naysayers about climate change too seriously, we cannot reject every line of metaphoric inquiry. Scientific metaphors, like any other, are not entirely open to choice, because there are external constraints on their aptness. Although I contend that there is greater flexibility for choosing scientific metaphors than scientists might assume, there are undoubtedly contexts in which certain metaphors simply do not work. Nonetheless, citizen involvement can help ensure that scientists are not closed to alternatives, as these citizens may well make novel suggestions and have sensitivities and thus ask questions that scientists do not, questions that may have significance for both epistemic fertility and social consequence.

This discussion may play its greatest role by simply improving awareness of issues associated with selecting scientific metaphors. This could contribute to grassroots inclusion of more diverse perspectives in their selection. Rather than scientists themselves making such decisions, they may informally consult with a broader range of actors than other scientists. I have suggested some axes for their deliberations, but others would evolve over time and arise in the context of specific conversations about particular metaphors in particular contexts. Thus, we may not require formalized methods for making such decisions. It seems highly unlikely that we would have ethics committees representing humanity screen potential metaphors before they can even be used in a lab—as if they are poisons. We all encounter enough bureaucracy as it is, and such processes might inhibit scientific creativity, even though, as I have argued, such creativity should not come without social responsibility. As one example of a more reasonable method, citizens might be involved with extended peer review of submitted papers proposing novel metaphors.[10] This would have been a possibility for the papers that coined the two metaphors, DNA barcoding and invasional meltdown, which I discussed earlier. This method may work for relatively discrete introductions, but there are other contexts in which new metaphors arise more subtly. Rather than simply assuming that dialogue is unnecessary for new metaphors, we require a better sense of when it is.

As is the case with many environmental issues, such evaluations are fraught with uncertainty. That is their nature. This circumstance, combined with how these metaphors link facts with values and science with society, is the very reason that we require new models for evaluating them. Although such uncertainty prevents us from predicting the effects of a metaphor with any accuracy, we can begin to attain measures of interpretation and local effects. A major challenge is that such interpretations will be subject to tremendous temporal and spatial variability. Not only will the meanings of metaphors change over time, but people from different places are likely to respond to them differently. Deliberative processes generally operate best within a local context, yet environmental metaphors know no such bounds. The metaphor of DNA barcoding, released in Guelph, Ontario, in 2003, has now traveled around the world. In some places, its connection to its consumeristic source might be tenuous, whereas in others the metaphor might solidify a consumeristic way of viewing the natural world that might otherwise remain incipient. We cannot foresee all such transformations, and thus we would be limited to local and contextual evaluation and only sometimes capable of more widespread forecasting. Such forecasting will require an awareness of how our Western worldview harbors assumptions that are not always in the interests of sustainability.

Assessing Public Values

One element of early participation in the selection of environmental metaphors is better assessment of their consistency with public values. By addressing this question, we would be better placed to foresee how people would respond to metaphors and perhaps even whether they are likely to be adopted. In a review paper in the *Proceedings of the National Academy of Science,* Michael Novacek, a paleontologist at the American Museum of Natural History, declared that we need a clearer and more

compelling message to engage people in "committed stewardship of biodiversity." To attain this goal, he proposed that we
require "(i) improved understanding of the diverse public audiences we are trying to reach, (ii) crafting of the messages
suitable for those diverse audiences, and (iii) enhancements
of the mechanisms for delivering those messages and eliciting
engagement," and he offered recommendations for each of these
in the context of biodiversity.[11] It is just this type of approach
that we require for environmental metaphors more generally.

Novacek's proposal mirrors a fourth type of expertise,
public expertise, which involves assessing public values.[12] If we
don't understand these, then we are unlikely to adopt appropriate metaphors. We require metaphors that are both consistent
with our values and objectives and sufficiently consonant with
those of the receiving audience. By assessing how they are likely
to be interpreted in particular contexts, we may forestall potential misinterpretations that could cause people to reject a
conception they might otherwise accept. We should rely not
merely on scientists' intuitions in such matters, but on the best
available information. It is an empirical question, for example,
whether DNA barcoding leads people to appreciate organisms
more or less. It is similarly an empirical question whether the
metaphor of an invasional meltdown causes action in response
to invasive species.

To address such questions, we can learn from marketing
firms that provide information about an audience so that advertisers can be more strategic. In particular, such research can
parse an audience to understand the influence of background
knowledge, cultural context, and other factors on people's values and on their receptiveness to alternative frames and metaphors. We have a variety of methods at our disposal for such
assessments, ranging from more qualitative focus groups to

more quantitative survey methods and even psychological testing. Ideally, we would merge the results obtained from multiple methods, for example by using focus groups to obtain a sense of local issues, if any, connected with a metaphor, which might justify in-depth interviews or a more geographically widespread survey (like the ones presented in earlier chapters). Focus groups might also be used to suggest and evaluate alternative metaphors that could then be assessed more quantitatively to understand the values they activate in diverse respondents. The results of such inquiry will be quite contextual; a particular metaphor may have a fairly predictable effect, but there will always be exceptions.

Through these methods, we seek a better understanding of how people respond to particular metaphors. Once the futility of a discrete boundary between science and society is recognized, we will be more cognizant that environmental metaphors can be configured in the context of social marketing, where a primary objective is to listen to the public and learn from it. By conducting research on our audiences, we can tailor messages to them. For example, to produce its public report on evolution, *Science, Evolution, and Creationism,* the U.S. National Academies used focus groups and survey research to determine how to frame its message more effectively in relation to audience values. Surprisingly, this research found that people were persuaded not by legal arguments as much as by discussion of the benefits of evolutionary science for societal progress and its compatibility with religious views.[13]

As another example, briefly consider invasion biology. Appropriate studies might find that there is widespread support for an extermination campaign for a certain species in a particular demographic group but not in another. It is unlikely that advertising a war against invasive species, something we see all the time, would be effective with the latter audience. If that group's support is required, research could be conducted to determine where its values are consonant with the eradication campaign. The objective here is not to dupe the audience, as some advertising does, but to seek a form of communication that engages shared values. Until we conduct such research on public values with regard to invasive species, we will fail to understand the variation in people's sensitivity to militaristic metaphors, which is likely to depend on such factors as the severity of the invasion, public awareness of it, and even the extent to which local people are comfortable with militarism. In the process, we can also evaluate alternatives. Gurevitch, for example, suggests two alternatives to the metaphor of invasional meltdown in a single sentence: "Runaway positive feedbacks in a system create 'snowball' effects in which a phenomenon builds on itself in an accelerating fashion, becoming unstoppable."[14] We have very little understanding of how people would respond to either of these metaphors, *positive feedback* or *snowball.*

The objective of these methods is not to emulate social science as a new form of top-down expertise to replace natural scientists, not least because we can only imperfectly predict the effect of a metaphor. There will also be inevitable trade-offs with the metaphors we select. Psychological research, for example, has shown that *climate change* is associated with scientific discourse and elicits less concern from a sample of British residents than *global warming.*[15] The latter, however, contributes to the mistaken interpretation that the issues are predominately heat-related. In instances such as this audiences may need to be segmented, though in some cases there will be an inevitable trade-off between scientific precision and public engagement.

Unfortunately, there are relatively few empirical studies of the efficacy of particular metaphors for us to go by. Some

of the best work has been done by a professor of speech communication at the University of Georgia, Celeste Condit, who
has examined the extent to which shifts in genetic discourse
contribute to social change. She noted that although it is quite
difficult to predict such effects, the "desire to anticipate future
developments is probably laudable and irrepressible."[16] Nonetheless, she questioned the work of critics who assume that
genetic discourse, in particular the blueprint metaphor (which
likens genes to the architectural blueprint for an organism),
would be interpreted in a determinist and essentialist way and
thus contribute to a more discriminatory society, because those
with certain genetic conditions could be shunned. Metaphors
can be attributed too much causal power and some of her work
has convincingly shown that the blueprint metaphor is not necessarily interpreted deterministically. She also demonstrated
that although the shift from blueprint to recipe metaphors for
the genome recommended by some critics may have taken
place, it has not had the expected effect. One might expect that
alternative recipe metaphors would communicate the context-
dependent nature of development and the intermixing of gene
and environment, but she relied on a number of methods, including audience and persuasion studies, to demonstrate that
this may not have happened. She suggested that the critics may
have predicted incorrectly because of their implicit egalitarianism and assumption that metaphors are stable, when they are
quite dependent on context.

Nonetheless, Condit went on to demonstrate that the
critics were in a sense correct, but not for the reasons they
thought. She specifically compared how the discourse of genetics operates for poor people in a public hospital compared to
middle- and upper-class people in a university hospital, showing that it is of less use to the former because they gain little

understanding from it. She concluded that the metaphor may not lead to discrimination per se, yet genetic discourse can lead to it in an institutional context to the extent that it closes down meaning. She found that the hospital as an institution created physical and bureaucratic structures around the discourse of genetics and the need for testing, even though this discourse served little use for many patients. In such a manner, a metaphor can become established in actual practice, making it that much more difficult to transform.

We know relatively little about the effects of specific metaphors on environmental practice, let alone on the contrast between alternative metaphors. Witness Garrett Hardin's metaphor, "the tragedy of the commons," which recognizes the tendency for individual persons to act in their own self-interest in the depletion of shared resources even when doing so is not in the long-term interest of the group. Although social psychologists initiated a flurry of activity looking into whether the metaphor would apply, focusing on the conditions under which people cooperate, there has been much less study of the effect of the metaphor on practice. Experiments by three American psychologists in the early 1990s, however, demonstrated that the commons metaphor is memorable and encouraged a specific environmental behavior, less littering after a movie.[17] Though not compelling in itself, such research provides further evidence for the link between metaphor and practice.

The values I am discussing may be too green and too radical, so it's possible that people would be more responsive to the very metaphoric values I critique. Stated another way, perhaps we should sell to the lowest common denominator in the prevailing market. I prefer a vision of long-term engagement, as I will discuss in the next section, but it may well be reasonable to adopt different strategies for short- versus long-term adjustment of metaphoric values. Two psychologists, for example, have shown that Americans tend to have self-enhancing (egoistic) values, whereas pro-environmental behavior is associated with self-transcendent values.[18] This means that framing environmental messages to incite guilt and encourage sacrifice, as occurs commonly, will be ineffective because these approaches will be inconsistent with the values most Americans hold. They question, however, whether it is appropriate to pander to these values by reframing to appeal to them or whether it would be better to try to change them. Although they contend that the latter must be our long-term objective, it might be counterproductive in the short term, so they recommend gradually shifting the frame.

We can only imperfectly predict the consequences of metaphoric choices because of their complexity. Complexity is not the same as being complicated; in principle you could take apart something complicated, such as a jet, and put it back together. In contrast, complex things have emergent properties. Some would argue we just need better positivistic understanding or better tools to solve the problem, but Paul Cilliers, a philosopher at South Africa's University of Stellenbosch, instead has proposed from a deconstructive point of view that we cannot know complex things in their entirety.[19] A major consequence is that our knowledge of complex things is by definition flawed, because we have to leave things out with reductive framing strategies. Reduction is both flawed and risky. It may seem that what we leave out is trivial, but these systems are often nonlinear, so we remain uncertain of the effects of our omissions in both space and time. In the current context, metaphors both highlight and hide, and what they hide may be critical. Uncertainty about their effects is not sufficient reason to forsake public involvement, for uncertainty motivated Funtowicz and

Ravetz to claim that we require postnormal risk assessment that includes diverse perspectives. An ethics of deconstruction recognizes that we need to do something and be critical of acting at the same time. Thus, we have to be more careful and have greater humility in the face of ever-present complexity. In other words, the web reminds us here that we are making a connection to something else with our metaphor, though there are other connections, spreading in myriad directions, some known and many unknown. So we must tread lightly.

The Evolution of a Learning Society

The foregoing discussion has dealt mainly with features of public participation in the introduction of novel metaphors. There is evidence that scientists are amenable to including lay knowledge in research in this way, but they have greater concerns about lay consumption of knowledge. Notwithstanding my rejection of the deficit model (explained in chapter 1), education certainly plays a role in the interpretation of metaphor, though perhaps not as much as we might at first think. For example, the metaphor *balance of nature* is part of vernacular understanding of ecology, whereas ecologists largely discount it. A psychologist and a biologist teamed up to survey both biology and nonscience majors, before and after ecology instruction, to examine their understanding of the balance of nature. They found that students largely misunderstood it, interpreting it in miscellaneous ways in accord with their everyday understanding of its connotations, including a circular confusion of cause and effect. Perhaps more worrisome, the researchers found that teaching had little effect on this result. They suggest that everyday meanings and understanding may inhibit learning about ecological concepts, and that this can have negative implications for policy—in this case to the extent that people assume that ecological systems are orderly and predictable. In concluding, they speculated "that the lack of a fixed meaning for the balance of nature term could lead to problems in education, public policy, and the transmission of ecological concepts to the general public."[20]

This conclusion, however, presumes that the solution lies in fixing meanings so that nonscientists will better understand what scientists mean. This can sometimes be successful. The sociologist José Julián López, for example, demonstrated how the polyvalence of the metaphor "nucleotide-bases-as-musical-notes that produce the 'music of life'" was stabilized in a Canadian museum exhibition, The Geee! in Genome. The displays were set up to activate certain semantic interpretations rather than others "through the inclusion, in the texts, of terms or signifiers and statements that denote or connote the preferred target meanings." Nonetheless, he observed that meanings can be closed down, but polyvalence cannot be entirely neutralized. In fact, in a review of challenges to biologists' attempts to stabilize meanings, the biologist Karen Hodges of the University of British Columbia concluded, "The major difficulty with such reform efforts is that language seldom changes by prescription, as shown by over 350 years of failed attempts."[21] As she explained, the major reasons for such failure are that most people will be unaware of such proposed revisions, they will continue to encounter the older meanings, and old meanings will adhere to new ones.

Scientists may assume that we can control resonance by better educating people if they misunderstand a term. Empirical evidence suggests otherwise, however. In another study, Condit and her colleagues used three methodologies to investigate the public perception of the term *mutation*. They found that people generally understood its genetic meaning, but that they gave it a negative connotation associated with science fiction movies and ideas about the effects of radiation. Despite the prevalence of this term, the investigators thus recommended that public health professionals avoid using it because it could contribute to the stigma disease carriers already bear. Some might instead propose educating people about its technical meaning, yet they emphatically concluded: "We do not see this as achievable. Even if it were possible to achieve much greater familiarity with the technical concepts of mutation, this would not necessarily erase the emotional associations. Therefore, public health personnel will need to adapt to the public understanding when communicating with the public."[22] Such metaphoric meanings may be intransigent to change through education.

Reflections on metaphoric stability might give rise to new foci for environmental education. For example, environmental scientists tend to think of metaphor with a standard (predicative) account, their role being to transmit one or more true statements. Their role in a contrasting (enactive) account has as much to do with their open-endedness and potential creativity, which directly counters the conduit model. John Foster, a philosopher at Lancaster University, proposed that the enactive account of metaphors is the appropriate view and that by adopting it society would be better prepared to engage with metaphors. Placing this discussion in the context of thinking about natural capital as a means toward sustainability, he claimed, "'How far will thinking of this *as* capital take us, and in what directions?' is always the crucial, open-ended question."[23]

In the learning society that Foster envisions, teachers would seek to engage their students with polysemy rather than attempting to control it by teaching only the preferred meaning of a metaphor. Scientific metaphors originate in society, so they continue to draw on ideas and emotions with which students are familiar in their everyday lives. It follows that students will interpret new concepts in accord with their worldview, which is unlikely to be shifted by a one-term course. In this context, it might be more appropriate for students to analyze diverse metaphors so that they can obtain a more complex view.[24] Their reflections on these metaphors would richly assist in their becoming engaged citizens, a vision that contrasts markedly with the idea that the purpose of science communication is simply to communicate factual information.

If the point of metaphor is not just to communicate truth predicates, but to help us recognize that the world both is and is not as we see it, then we will develop a better appreciation for the complexities involved in environmental decision making. For example, if I speak of invasive species, I have to understand that they are *not* invaders, but an expression of our very selves. If we treat a DNA sequence as a barcode, we may obtain a useful tool, yet we simultaneously treat something that is not a barcode as if it were one. Although the predicative view would specify how these sequences are like barcodes, there is more to it than that. Barcoding elicits tensions concerning such issues as whether genetic sequences are like this or not. We can certainly conceptualize them this way, although there is a significant way in which these sequences are not like barcodes, and their referents, species, are not like barcoded objects. It is not just a matter of getting to the right answer and then stopping with metaphor, for metaphors' meanings need to be interpreted and reinterpreted. As we become more aware of multiple interpretations of a given metaphor (or even of alternative metaphors), the necessity for social engagement will become increasingly clear.

We can now return to the discussion of Gaia in chapter 1.

The Earth both is and is not like a living being, and we would benefit from living with the paradox of both of these views rather than gravitating to either extreme. The metaphor would then allow us to begin to query the dominance of the mechanistic paradigm. Gaia is a positive form of personification that focuses on our relationship with our home planet. If the Earth is a being, it is like us, and we are not only from it but also dependent on and related to it—in fact, part of it.[25] Scientists might consider this too wishy-washy an approach, and it took decades for them to accept the general idea on epistemic grounds. Interestingly, it was seen as too romantic, but the terms *barcoding* and *meltdown* are not seen as too unromantic. The "is not" of Gaia has been judged more harshly than that of these more recent metaphors, probably because they accord with pervasive, unquestioned values. Gaia takes on the greater challenge of shifting our worldview, helping us realize that there is something valid in that "is not."

The objective here, idealistic as it may seem, is to improve democracies by encouraging people to be more engaged in all aspects of their lives. Many would argue that scientific language is too complex to be understood by most people. But this presumption is just as likely an excuse to maintain power relations as they exist today. As Norman Fairclough, a linguist at Lancaster University who helped found critical discourse analysis, observed, "People cannot be effective citizens in a democratic society if their education cuts them off from critical consciousness of key elements within their physical or social environment."[26] Engaged citizens are entitled to critical language awareness. At the very least, I hope, this book contributes to a greater consciousness of metaphors and their role at the interface between science and society. In the process, citizens would become more aware that science is a self-adjusting way of exploring the world, rather than an accumulation of authoritative facts.

The message here echoes a famous speech given by Robert Frost before the Alumni Council of Amherst College on November 15, 1930. He spoke about the fundamental role of poetry in education, specifically insisting on the importance of metaphor as a means to critically evaluate one's place in the world. Although he excluded scientific thinking from metaphorical thought, he nonetheless noted, "Unless you are at home in the metaphor, unless you have had your proper poetical education in the metaphor, you are not safe anywhere. . . . You don't know how far you may expect to ride it and when it may break down with you. You are not safe in science; you are not safe in history." He provided the specific example of someone who attempted to defend the statement "the universe is a machine," though that person was eventually forced to acknowledge under Frost's tutelage that, unlike other machines, it doesn't have "a pedal for the foot, or a lever for the hand, or a button for the finger." Accordingly, "it is different from a machine." But Frost observed, "He wanted to go just that far with that metaphor and no further. And so do we all. All metaphor breaks down somewhere. That is the beauty of it. It is touch and go with the metaphor, and until you have lived with it long enough you don't know when it is going. You don't know how much you can get out of it and when it will cease to yield. It is a very living thing. It is as life itself." And, of course, even this analogy might lead to reflexive queries about the ways that language both is and is not like a living being.[27]

The Emergence of Ecological Ethics

Not only the public but also scientists themselves have responsibility for their metaphors. When scientists use an emotive
metaphor such as barcoding or meltdown, they are drawing
attention to a phenomenon. In the process, if their sales pitch
has been successful, they may obtain public funds for their research. The question, however, is whether this is appropriate,
given that such promotional activity may conflict with their
role in society. In particular, this promotion relies on values,
whether implicit or explicit, that not only need to be discussed
in a broader public process of deliberation, but also need to be
considered by scientists themselves. Environmental scientists
will continue to create new metaphors and to draw on ordinary
understanding as they do so, yet we can only hope that they will
increasingly reflect on the social dimensions of their choices,
especially given the authority of their pronouncements. To the
extent that they want to communicate with nonscientists and
to conduct research that bears on the human relationship to
ecological systems, reflection on their metaphors—at all stages
from their inception through their demise—becomes essential.

It might seem that I am overstating the significance of
these metaphors and particularly the role of scientists. Of all
the cultural forces relevant to bringing about environmental
reform, am I not overstating the role and importance of science,
compared to, say, religion and politics? Who will stop multi-
national companies from promoting whatever metaphors or
images they want through their advertising? Why should scientists be any more responsible for the spread of these metaphors
than people who use them widely in day-to-day discourse?
What about the media? And what about the military men who
promote war? These are appropriate questions, but they do not

excuse scientists from metaphoric responsibility. Science plays a significant role in decisions about the environment, so if environmental scientists want to be leaders in effecting change, they must become conversant in such issues at the science-society interface.

Some scientists may claim that it is difficult enough to find an apt metaphor without having to evaluate its social effects, yet I have argued that this reflects an untenable disengagement from society. There are always alternative ways to describe a phenomenon, and the ready availability of competitive, consumeristic, and militaristic sources, for example, is no excuse for a lack of careful thought and social engagement. Some might wish instead to rely on an evolutionary epistemology that winnows metaphors over time, yet Thomas Kuhn has reminded us, "A paradigm can, for that matter, even insulate the community from those socially important problems that are not reducible to the puzzle form, because they cannot be stated in terms of the conceptual and instrumental tools the paradigm supplies."[28] In the realm of sustainability, scientists may work within a narrow paradigm formulated by a few metaphors that are inconsistent with their broader intentions. We need to bring social dimensions to the fore before our metaphoric choices become entrenched.

In their review of alternative metaphors in social dilemma research, for example, three psychologists concluded that the triple metaphor of a *prisoner's dilemma game* has become "so deeply entrenched in the scientific lexicon that most scientists are largely unconscious of its metaphorical origins," despite some of its limitations when compared to the other forty-five metaphors that had been used in the previous half century.[29] They also reviewed the importance of these alternatives to environmental sustainability. This does not mean that such considerations have to drive ecological research; rather, we need to strive for language that best meets multiple demands in the face of uncertainty about potential efficacy. I have argued that we need to rectify the existing imbalance between evaluation of the epistemic and the social dimensions of metaphors in environmental science: the former is not perfectly predictable, either.

With such reflections, I propose that we need to broaden the usual ethical requirements of science, which extend mainly to commitments to truth telling and avoiding plagiarism and undue bias. Other scholars propose that conservation scientists need to engage more directly with ethical theory. Ben Minteer, a philosopher from Arizona State University, and his colleagues provide direction here in proposing a new area of inquiry, ecological ethics, which aims to assist ecologists with the ethical issues encountered in their work. Their proposal is germane to consideration of environmental metaphors: "Ecological ethicists have a central and substantial responsibility in a world where democracy is viewed and valued as the ideal. Yet the potential for manipulated opinion and discourse and the subversion of the decision-making process, as opposed to informed public debate and decision-making, are an ever present danger that must be explicitly recognized and addressed."[30] The case studies herein have shown that this is a particular danger with ecological metaphors. Environmental scientists can neither control established metaphors nor take into account every potential implication of novel ones that they propose, but they nonetheless need to give them greater attention in the name of a socially responsible biology.

The evidence presented in the previous two case studies suggests that environmental scientists assume they know what they are doing with their metaphors and that their choices are in everyone's best interest. They have not necessarily been trained to think carefully about their metaphors, however, and might thus have unexamined positivistic assumptions about them. For example, they may assume that they are mere rhetoric that has little to do with the things in themselves; they may not recognize how they entangle epistemology and rhetoric. They may also be relatively unaware of the value implications of their metaphoric choices, even with so-called dead metaphors.

Environmental scientists may also have little opportunity to develop an appreciation for the power of language. To develop such awareness, there is an educational movement that seeks to bring a greater consciousness of the process of language and how it works. Children benefit from learning additional languages, but we also need to encourage them to develop a critical view of language, how it works, and what it hides. Similarly, scientists' training may cover the epistemic precision of language, but they are less likely to learn to reflect on it more broadly (notwithstanding many individual exceptions). As a result, problematic patterns show up repeatedly when scientific language is subjected to ethical and critical scrutiny.[31] In particular, there may be an implicit hierarchy founded on the deficit model, whereby scientists are knowledgeable and the public simply requires adequate education. Such a view of language maintains existing power structures and limits our capacity to develop more democratic science-society relations.

Given the foregoing, I encourage environmental scientists at all stages of their careers to incorporate a modicum of ethical training, not only to raise questions about metaphors, but to provide a broader social context for their inquiries. We might imagine, for example, a code of conduct for the use of metaphor. Such a code would have to be flexible and seek an evolving middle ground between the extremes of objectification and catastrophism. If we think of our metaphors as flowers in a garden, for example, we could imagine that garden as being very tidy and trimmed or quite disorderly.[32] The purpose of a code is not to obtain some ideal of orderliness in the garden, but to maintain people's awareness of the need for some level of order—or at the very least to encourage discussion of how much order we would like to have.

A first element might simply parallel the Hippocratic oath: "When I use a metaphor, I will take into account how metaphors can cause harm and thus seek ones that seem least likely to do so." Another might be, "I will not promote my work with a metaphor until there is a relative consensus on the validity of the scientific results." Although there may be scientists who would release a new metaphor without input because they wager that it would be a successful meme that would catapult their careers, ideas, and research funding, we can only hope that both other scientists and the public would become more sensitive to such appeals in the future.

Part of this code might address the use of metaphors in education. Environmental scientists who act as educators might dismiss students' detection of metaphoric resonance as misunderstanding. As I mentioned earlier, however, such perceptions might derive from entire interpretive frameworks. Thus, education becomes not so much a means to inform people as an exercise in "foreign affairs." Educators can strike a balance between stifling political correctness and nonchalant insensitivity by discussing with students the benefits and limitations of various metaphors through the window of their own life experiences. In the end, this could lead to better learning. And toward that end, science teachers might benefit from considering alternative metaphors for pedagogy itself, especially ones founded on alternatives to the conduit and deficit models, which imply simply filling students' minds with information.[33]

Environmental educators might also keep in mind that counterproductive metaphors reinforce themselves in part through education to the extent that it is a conservative force. Chet Bowers, for example, raised a concern that "too often environmental education is a form of socialization to the eco-management way of thinking that is predicated on the root metaphors of anthropocentrism, subjective/rational individualism, and economism."[34] Many environmental educators were trained as scientists, and they pass along their worldview, with both its strengths and its limitations. Bowers is particularly concerned about the tendency to emphasize the benefits of science and technology without a balanced treatment of how they have contributed to many of our problems. This emphasis tends to downplay the necessity for more humanistic solutions, including the preservation of cultures, as we instead await techno-scientific solutions (as in the case of DNA barcoding).

A metaphoric code would also have to address the question of advocacy. There is ongoing debate about this issue that I cannot aim to resolve here.[35] A commonly proposed solution is that advocate-scientists should wear two hats, one when speaking as scientists and another when speaking as citizens. The former concerns the empirical facts of the matter, whereas the latter relates to their personal positions. There is value in this ideal, and scientists should undoubtedly try to clarify this distinction whenever possible.

Nonetheless, there are problems with the two-hat model when metaphors are involved. For example, consider the recent typology developed by two German scholars for the diverse meanings of *resilience*. Along with numerous other scholars, Fridolin Brand and Kurt Jax suggest that this metaphor is a critical one for thinking about sustainability, even though its meanings range from descriptive-ecological to quite normative ones.

The latter allow resilience to act as a useful boundary object for communication, but Brand and Jax argue that as this usage continues to increase, the value of resilience for scientific inquiry simultaneously declines. As a solution, they propose a variant of the two-hat model, where resilience is operationalized as "ecological resilience/ecosystem resilience" within ecological science, at the same time being dubbed "social-ecological resilience" when used as "a vague and malleable concept that is used as a transdisciplinary approach to analyze social-ecological systems."[36] Despite their good intentions, this book has demonstrated how metaphors such as resilience will foil the two-hat model in practice. We cannot control the resonance of resilience or related terms. In the current context, I suggest that we do not even understand its contemporary resonance and thus whether it is useful for communicating ideas more broadly. Even if such metaphors need to be clarified in science, as Brand and Jax suggest, this will probably have little effect on their meaning in public. The popular meaning will prevail because such metaphors carry value-laden messages.

The two-hat metaphor is telling in itself. The person underneath the two hats is the same, which demonstrates the actual challenge of separating these two perspectives. To think that one can do so merely embodies the fact-value and science-society dichotomies within a single person. Regardless, ordinary citizens will continue to detect values in science even when scientists think they are being objective and wearing their scientist hats. Scientists who take on this role speak as pure scientists, but they effectively act as stealth issue advocates. Instead, scientists need to learn to articulate their values even when they think they don't have them. This opens up a two-way conversation that is essential both to good communication and to the interpretation of metaphors.

VIII
Wisdom and Metaphor

*And the animals, instinctively, have already noticed
that we aren't really at home in our talked-about world.*
—*Rainer Marie Rilke, "The First Elegy,"* Duino Elegies

*[The world] was once full of Gods; it then became a
drab material world; and it can be changed again, if its
inhabitants have the determination, the intelligence,
and the heart to take the necessary steps.*
—*Paul Feyerabend,* Conquest of Abundance

E ven though sustainability is by no means assured, we
nonetheless seek a more enduring relation between
ourselves and the world in which we live. Environmental science certainly has a role to play here, but
these stories of metaphor have shown it is much more likely
to do so by recognizing and engaging with its broader social
context. The conscious use of feedback metaphors—and engagement with their values—is one way to enliven a form of
science that can better help formulate an alternative future. This
form of science is more widely distributed in society, and thus
everyday citizens have a key role to play.

It may seem that I've focused too much on weighing the
benefits and costs of individual metaphors. Although that is
necessary, I think that the deeper lesson of this book is how
metaphors can prompt dialogue between people with different
perspectives. Sustainability will require the wisdom to be able
to relate to one another. And alternative metaphors can help,
reminding us that no single metaphor can capture a phenomenon in its entirety, because every metaphor highlights certain elements while veiling others. Diverse metaphors provide
varied perspectives, and thus differing ways of experiencing
or relating to phenomena. I am reminded of the Indian fable
about the blind men and the elephant, each of them touching
a different part and interpreting "elephant" accordingly—and
incorrectly—because each of their views is partial. If the men
had instead communicated their perspectives with one another,
they would have arrived at a more complete image of the crea-
ture in question. The eminent psychologist William James promoted the benefits of holding diverse metaphoric approaches
in mind during scientific studies, a pattern he himself evinced
in the "ensemble" of metaphors he used to characterize human
psychology.[1] Otherwise, focusing on only one metaphor can

actually divert us from alternative perspectives: if we think an elephant is characterized and therefore defined by rough skin, we may never discover its snout.

Metaphoric plurality is important in the epistemic realm, but I have shown that it is even more critical at the intersection of science and society in the context of seeking sustainability. Here we will benefit from the playful interaction among points of view represented by different metaphors that codify values representing the interests of diverse stakeholders. The contact and interplay between different metaphors bring values into the open for discussion, which can reduce conflict and contribute to more democratic environmental decision making. Alternative metaphors lead to different ways of seeing social problems. Often such problems result from a conflictual situation in which "the adversaries do not disagree about the facts; they simply turn their attention to *different* facts." In such a situation, stakeholders might explore the interface between existing metaphors through a process of "frame restructuring" that can give rise to an entirely new way of seeing.[2]

When a scientist or group of scientists commit to one metaphor, they are likely to be seen as issue advocates, when a better role, in complex environmental issues, is for them to be honest brokers of policy alternatives. Rather than closing down the options, they can then increase them, providing the full range of policy possibilities. Expert and nonexpert alike are thus encouraged to attach less truth to their metaphors, which can facilitate more engaged discussion, learning, and problem solving. We can expand a stale slate of candidate source domains through open deliberation among those with differing backgrounds, both expert and nonexpert, from the outset. This is not the same as replacing one privileged discourse with another, even if it is thought to be more green. In his book on the debate over genetic modification (GM), *Genetically Modified Language,* Guy Cook likened this distinction to that between a monoculture and a mixed crop. Those who promote monocultures of GM plants tend to rely on an analogous monoculture of scientific data and language for resolving the GM controversy. In contrast, he endorses the benefits of "a mixed crop of varying perspectives upon the problem."[3] Science provides a crucial perspective, certainly, but it is enmeshed in a societal context that must be considered more explicitly. Our approach to language and to land may well mirror one another, so we might cultivate wisdom through metaphor, a wisdom born of the empathy that arises through a growing recognition of other perspectives, which both increases the richness of our own view and our ability to engage in discussions with others.

Metaphors not only connect us with one another. They may also act to renew our relation with the natural world. It is particularly critical that we select metaphors that encourage our children to develop an intimate and embodied relation with it (Figure 16). Fear-based appeals about invasive species can be counterproductive, and perhaps even more so for sensitive children who may already have enough reasons to be afraid of nature and who often grow up in urban spaces where there are many invasive species. Similarly, we need to better understand whether barcoding will increase kids' involvement with nature—and not just assume that it will. In a world faced with climate change and human population growth, invasive species and biodiversity loss, we require metaphors that are hopeful as opposed to catastrophic, that are inclusive rather than deceptive, and that thereby bring us closer to the world rather than separating us from it.

In our search for sustainable metaphors, we inhabit the

Figure 16. A child interacting with another species.
Photo courtesy of James Kamstra.

interface between science and poetry. Wendell Berry once observed, "Applying knowledge—scientific or otherwise—is an art. An artist is somebody who knows what to put where, and when to put it." We need to place our metaphors with care because they tend to reinforce preexistent ways of thinking. And to do so, we must continually heed the interdependence of facts and values and thus science and society in our metaphors. This will contribute to the adoption of metaphors that will be more conducive to the conceptual shift we require for sustainability. We cannot rely just on conventional metaphors because we will occasionally require more novel, poetic ones that help us see the limits of ordinary language use. They may also help express the depths of our fears and sorrows, which may be constrained by normal, utilitarian, and problem-solving language; as Walt Whitman, in "Song of Myself," expressed it: "What living and buried speech is always vibrating here, what howls restrain'd by decorum." Part of what a poet does is to come up with novel language that reveals something that was previously unnoticed.[4] Such metaphors can help enliven environmental science, especially when they inspire love for local places that people care about.

Perhaps one of the more profound reminders from these reflections is that life is not language, not even the language of environmental science. There is something not captured by particular words and metaphors, perhaps what the mystics call "suchness." Its recognition can be invigorating, as we can then be in touch with the actual freshness of experience that is not mediated by language.[5] Feyerabend calls the loss of this experience the "conquest of abundance," whereby science and other such systems reduce the abundance of life around us into reductive and ultimately false systems that are given more importance than our holistic experience. The challenge is to keep returning to that experience, whether by art or by science, to bring back signs across the abyss and form them into words that inhabit us and allow us to communicate with others. It is not to convince others, for none of us understands entirely what this is that we inhabit, so we must share our perspectives to obtain a more expansive sense of the whole.

Nonetheless, I have paradoxically suggested that we need to connect with the world through language, an intermediary that many would claim fundamentally separates us. By simply attending to our metaphoric choices and reflecting on them, however, we recall that there is life beyond language and we thus acquire greater humility and responsibility for the words we put to our experience of the world. I can return to the ramblings in the natural world with which I began, falling in love with it, and thus wanting to express that love—along with everyone else—in the most apt words I can.

I turn away from these metaphorical musings and walk through the fallen leaves beneath towering sugar maple trees, their colors rustling as I pass. The bumblebees and spring wild-flowers are long gone, though wands of zigzag goldenrod still bring color to the greens and browns of the understory. I reach a wall of limestone, where shades of vibrant green texture reveal three species of moss. I could imagine asking questions about how they coexist here, perhaps somewhat banal questions about whether they compete with one another, but I can never resist reaching into their cool and timeless embrace; I never tire of the wonder they bring—a wonder that I wish everyone could experience to understand why all of *this* is so wondrous. And one way we can communicate the meaning of such experiences is with metaphors that link science and the everyday, creating tremors in the metaphoric web we must initiate with care.

Notes

I
Metaphor and Sustainability

1. On the role of language in the construction of facts, see Fleck, *Genesis and Development of a Scientific Fact;* Gusfield, "Literary Rhetoric of Science"; Latour and Woolgar, "Laboratory Life"; Shapin and Schaffer, *Leviathan and the Air-Pump;* Bazerman, *Shaping Written Knowledge;* Gross, *Rhetoric of Science;* Myers, *Writing Biology;* Killingsworth and Palmer, *Ecospeak;* Fahnestock, *Rhetorical Figures in Science;* Beer, *Darwin's Plots.* On large-scale metaphors in the history of ecology, see Worster, *Nature's Economy;* Hagen, *Entangled Bank.*

2. Brown, *Making Truth,* 14; Gould, "Self-Help for a Hedgehog," 1021. The literature on metaphor is vast, so in part for this reason I will be drawing largely on relatively recent contributions. For an extensive bibliography of early work on metaphor, refer to Shibles, *Metaphor.* On the historical denigration of metaphor in science, see Ortony, "Metaphor"; Keller and Lloyd, "Introduction." Some classics from the extensive corpus concerning its importance in science include Black, *Models and Metaphors;* Schlanger, *Les métaphores de l'organisme;* Haraway, *Crystals, Fabrics, and Fields;* Boyd, "Metaphor and Theory Change"; Kuhn, "Metaphor in Science"; Hesse, *Revolutions and Reconstructions;* Keller, *Refiguring Life;* Bono, "Science, Discourse, and Literature"; Keller, *Making Sense of Life;* Baake, *Metaphor and Knowledge.* Nonetheless, scholars still debate whether metaphors are constitutive or just heuristic, that is, merely useful tools. More radically, linguists have recently critiqued the view that metaphors are parasitic on literal meaning on the grounds that the implicit distinction between literal and metaphorical meaning is unfounded; see, e.g., Rumelhart, "Some Problems with the Notion of Literal Meanings"; Hesse, "Cognitive Claims of Metaphor"; Gibbs, *Poetics of Mind;* Lakoff and Johnson, *Philosophy in the Flesh.*

3. Analogies, metaphors, and models are closely related to one another, and I won't belabor the extensive literature concerning the distinctions between them. Analogies are more explicit than metaphors and suggest that two things are similar in the same way as two other things. Metaphor is also closely related to other figures, particularly metonymy and synecdoche; see, e.g., Lakoff and Johnson, *Metaphors We Live By;* Barcelona, *Metaphor and Metonymy.*

4. Lakoff and Johnson, *Philosophy in the Flesh,* 22. The resulting view has been called experiential realism or embodied scientific realism. Though I will not explicitly adopt the conventions of cognitive linguistics in what follows (though see discussion of ecolinguistics in chapter 4), its insights underlie much of my thinking about metaphors and their significance.

5. Cuddington, "'Balance of Nature' Metaphor," 465. Note that *balance* in this discussion is technically an image schema; see, e.g., Gibbs, "Embodied Nature of Creative Cognition." For further discussion of the balance of nature and the shift to a nonequilibrium view, see Worster, "Order and Chaos"; Wu and Loucks, "Hierarchical Patch Dynamics"; Cooper, "Balance of Nature"; Lodge and Hamlin, *Religion and the New Ecology.* For a related discussion of force in natural science, see the perceptive commentary in Young, "Darwin's Metaphor and the Philosophy of Science," and the exchange between the evolutionary biologists Richard Dawkins and Stephen Jay Gould in Gould, "Self-Help for a Hedgehog."

6. For more on this cultural view of metaphor, which has developed partly in response to the perceived cognitive bias of Lakoff and Johnson, see Quinn, "Cultural Basis of Metaphor"; Gibbs, "Taking Metaphor Out of Our Heads"; Eubanks, *War of Words;* Frank et al., *Body, Language and Mind.*

7. Classic expositions of how language and metaphor interweave science and society include Stepan, "Race and Gender"; Kwa, "Representation of Nature"; Taylor, "Technocratic Optimism"; Bono, "Science, Discourse, and Literature"; Keller and Lloyd, "Introduction"; Maasen and Weingart, "Metaphors—Messengers of Meaning"; Maasen, Mendelsohn, and Weingart, *Biology as Society, Society as Biology;* Gaziano, "Ecological Metaphors"; Otis, *Membranes;* Rozzi, "Reciprocal Links"; Pickett and Cadenasso, "Ecosystem as a Multidimensional Concept"; Otis, "Metaphoric Circuit." I am not denying that within science language may be used in ways somewhat different from those in everyday contexts, but arguing that the use of everyday words in science tends to erase this distinction in practice. For further explanation of how the meaning of scientific metaphors may be opened and closed, see Knudsen, "Scientific Metaphors Going Public."

8. For more on metaphor and the process of conventionalization, see Schön, *Invention and the Evolution of Ideas;* Trim, *Metaphor Networks.* For

an intriguing analogy between the literalization of old metaphors and the formation of coral reefs, see Rorty, *Contingency, Irony, and Solidarity*, 16.

9. Harrington, "Metaphoric Connections," 359; Gilbert, "Metaphorical Structuring of Social Perceptions," 184. On performativity, see Austin, *How to Do Things with Words*, and on the performativity of metaphor, see Lenoir, *Inscribing Science*; Bono, "Metaphorics of Scientific Practice." On worldviews, see Smart, *Worldviews*. Note that the term world-"view," in its emphasis on the visual sense, may imply a separation of humanity from nature; see, e.g., Heidegger, *Question concerning Technology*, 133–134. On how metaphors contribute to a "form of life," see Wittgenstein, *Philosophical Investigations*. On the influence of medical metaphors on patients, see Martin, "Toward an Anthropology of Immunology"; Sontag, *Illness as Metaphor*; Kirmayer, "Body's Insistence on Meaning"; Low, "Embodied Metaphors"; Gwyn, "'Captain of My Own Ship'"; Rasmussen, "Poetic Truths and Clinical Reality." For classic views of how scientific metaphors influence human culture, see Turbayne, *Myth of Metaphor*; Schön, "Generative Metaphor"; Gilbert, "Metaphorical Structuring of Social Perceptions"; Lakoff and Johnson, *Metaphors We Live By*. For further discussion of how they affect our relationship to the natural world, see, e.g., Mills, "Metaphorical Vision"; Porteous, "Bodyscape"; Meisner, "Metaphors of Nature"; Philippon, *Conserving Words*; Verhagen, "Worldviews and Metaphors."

10. Lakoff and Johnson, *Metaphors We Live By*, 145. The examples of Time Is Money here are adapted from ibid., 7–8, which provides further discussion. On the commodification of time, also see Goatly, *Washing the Brain*.

11. On the former, see Lovelock, *Gaia*; on the latter, see Ward, *Spaceship Earth*; Fuller, *Operating Manual for Spaceship Earth*. For critical perspectives on the metaphor of spaceship Earth, see Edge, "Technological Metaphor and Social Control," and, on the metaphor of Gaia, see Murphy, "Sex-Typing the Planet"; Jelinski, "There Is No Mother Nature." On the positive view of Gaia presented here, see Midgley, *Earthly Realism*. On the role of the Gaia metaphor as a counter to the predominant mechanistic one, see Abram, "Mechanical and the Organic." For a comparison of three metaphorical models for the relation between humans and the ecosphere—predator-prey, tumor-host, and the preferred parasite-host—see Henrich, "Gaia Infiltrata."

12. See Hull et al., "Assumptions about Ecological Scale."

13. Schön, "Generative Metaphor," 138.

14. Writings on the epistemic limitations of particular ecological metaphors include Mikkelson, "Methods and Metaphors"; Hurlbert, "Functional Importance vs Keystoneness"; Slobodkin, "Good, the Bad and the Reified"; Cuddington, "'Balance of Nature' Metaphor"; Pickett and Cadenasso, "Ecosystem as a Multidimensional Concept"; Haila, "Conceptual Genealogy of Fragmentation."

15. Funtowicz and Ravetz, "Science for the Postnormal Age." Their term *postnormal* contrasts with the puzzle solving of Kuhnian normal science; see Kuhn, *Structure of Scientific Revolutions.* The waves of science-society interaction are discussed by Collins and Evans, "Third Wave of Science Studies." See the related discussion of mode-2 knowledge production in Gibbons et al., *New Production of Knowledge.* For an early critical perspective on the very notion of a divide between lay and expert knowledge, see Wynne, "May the Sheep Safely Graze?" For critiques and reviews of the linear model and arguments for more deliberative and democratic relations between environmental science and society, see Beck, "From Industrial Society to the Risk Society"; Dietz and Stern, "Science, Values, and Biodiversity"; Kleinman, "Beyond the Science Wars"; Jasanoff, "Technologies of Humility"; Robertson and Hull, "Public Ecology"; Kitcher, "Responsible Biology"; Latour, *Politics of Nature;* Wilsdon and Willis, *See through Science;* Wilsdon, Wynne, and Stilgoe, *Public Value of Science;* Pielke, *Honest Broker.* Statements by scientists themselves that promote greater societal engagement or the need for more interaction between natural and social science include Bazzaz et al., "Ecological Science and the Human Predicament"; Lubchenco, "Entering the Century of the Environment"; Bradshaw and Bekoff, "Ecology and Social Responsibility"; Balmford and Bond, "Trends in the State of Nature"; Palmer et al., "Ecological Science and Sustainability"; Uriarte et al., "Broader and More Inclusive Value System"; Foote, Krogman, and Spence, "Should Academics Advocate?"

16. See chapter 2 on the metaphoric web. The view of metaphor here draws largely on the interactive model of metaphor whereby the two referents interact with one another by drawing on "associated commonplaces," as Black called them, to produce new meaning; see Richards, *Philosophy of Rhetoric;* Black, *Models and Metaphors.* For a discussion of an alternative though still musical metaphor, harmonics, see Baake, *Metaphor and Knowledge,* a rhetorical analysis of metaphors employed at the Santa Fe Institute.

17. See chapter 6 on the metaphors of invasion biology. On racist and sexist metaphors in environmental science, see Zuk, "Feminism and the Study of Animal Behavior"; Janusz, "Feminism and Metaphor"; Herbers, "Watch Your Language!"; and the classic work by Haraway, *Primate Visions.* On the changing paradigm of the role of disturbance in ecological systems, see Worster, "Order and Chaos"; Lodge and Hamlin, *Religion and the New Ecology.* On the risks of communication, see Weingart, Engels, and Pansegrau, "Risks of Communication." On the evolutionary ecology of metaphor, see Nerlich, "Tracking the Fate."

18. On the deficit model, see Hilgartner, "Dominant View of Popularization"; Irwin and Wynne, *Misunderstanding Science?;* Gregory and Miller,

Science in Public; Locke, "Golem Science"; Ungar, "Knowledge, Ignorance and the Popular Culture."

19. Hilgartner, "Dominant View of Popularization," 528. Numerous studies have shown that scientists variably adjust their standards to give authority to science (versus nonscience) in different domains; see, e.g., Gieryn, *Cultural Boundaries of Science.* One of the best case studies of rebellion against normal science is an analysis of how AIDS patients called for a revision to drug-testing protocols that would otherwise withhold treatment from some of them; see Epstein, *Impure Science.*

20. See Reddy, "Conduit Metaphor." For an insightful discussion of the conduit metaphor in the context of science communication, see Weber and Word, "Communication Process." In a very different context, it has been argued that the notion of hereditary transmission gave rise to Crick's central dogma, which relies implicitly on a conduit model; Oyama, *Evolution's Eye,* 50–52.

21. Keller, "Language and Ideology in Evolutionary Theory," 91–92. This statement was made in the context of her discussion of competitive metaphors in ecology.

22. Nisbet and Mooney, "Framing Science." Also see the reply to their paper by Holland et al., "Letters." For further introduction to framing, see Entman, "Framing"; Gray, "Framing of Environmental Disputes"; Lakoff, *Don't Think of an Elephant!;* Beardsley, "Framing Biology"; Nisbet, "Ethics of Framing Science."

23. Numerous scholars have set a precedent for expanding our assessment of scientific metaphors into the social realm, in addition to the medical ones mentioned earlier. For example, there has been extensive critical assessment of whether code, blueprint, and related metaphors give an overly deterministic impression of the influence of one's genes and thus whether alternative metaphors might better reflect modern genomics; these include Avise, "Evolving Genomic Metaphors"; Nelkin, "Molecular Metaphors"; Nordgren, "Metaphors in Behavioral Genetics"; Turney, "Sociable Gene." I will refer to numerous examples from the environmental sciences throughout this work, including the one mentioned here: Pickett, Cadenasso, and Grove, "Resilient Cities." The closest parallel to the argument in this book of which I am aware is a paper on the ethical significance of language in pollution research by Elliott, "Ethical Significance of Language."

24. Goatly, *Washing the Brain,* provides a thorough introduction to the ideological effect of metaphors. On dominant versus outlaw discourse, see Ono and Sloop, *Shifting Borders;* on social disturbance, Philippon, *Conserving Words,* 5; on memes and cultural evolution, Ehrlich, "Intervening in Evolution" (a meme is a unit of cultural evolution, analogous to a gene in biological

evolution). For feminist writing on how metaphors can be a subversive force, see Janusz, "Feminism and Metaphor"; Rosner and Johnson, "Telling Stories." For additional expressions of the importance of language in shifting to sustainability, see Bowers, *Education, Cultural Myths, and the Ecological Crisis;* Oelschlaeger, *Caring for Creation,* 119.

25. Lakoff, *Don't Think of an Elephant!* Though this work has been widely adopted by liberals, note that it has also been quite critically reviewed, particularly for its reliance on an essentialist dichotomy between conservatives and liberals.

26. Quoted in Seed, "Beyond Anthropocentrism," 39. In this sense, my work is affiliated with a notion of "subversive" ecology promoted in the 1960s; see, e.g., Sears, "Ecology—A Subversive Subject"; Shepard and McKinley, *Subversive Science.*

27. Scholars calling for more value-laden environmental language include Ross et al., "Ecosystem Health Metaphor"; Westoby, "What Does 'Ecology' Mean?"; Norton, "Improving Ecological Communication"; Hull and Robertson, "Language of Nature Matters"; Ungar, "Knowledge, Ignorance and the Popular Culture"; Callicott, "Multicultural Environmental Ethics"; Trudgill, "Psychobiogeography"; Weber and Word, "Communication Process"; Hull et al., "Understandings of Environmental Quality." This may be seen as part of a broader social trend, in politics and elsewhere, to increasingly use everyday language in public contexts; see, e.g., Fairclough, *Critical Language Awareness,* 4; Blommaert and Bulcaen, "Critical Discourse Analysis," 453. For an interesting extension of this argument to the need for conservation scientists to adopt the language of social traditions, see van Houtan, "Conservation as Virtue."

28. See Lach et al., "Advocacy and Credibility of Ecological Scientists"; Wallington and Moore, "Ecology, Values, and Objectivity"; Lackey, "Science, Scientists, and Policy Advocacy"; Kincaid, Dupré, and Wylie, "Value-Free Science?"; Pielke, *Honest Broker;* Foote, Krogman, and Spence, "Should Academics Advocate?" For a recent review, which concludes that it is a question not of whether to advocate, but how much, see Nelson and Vucetich, "On Advocacy by Environmental Scientists."

29. Westoby, "What Does 'Ecology' Mean?"; "on whose shoulders" comes from Thomas, *Late Night Thoughts,* 155; Nabhan, *Cross-Pollinations,* 13, 71. I can claim to be only a novice poet, yet I nonetheless see the value of such cross-pollination and wish to nurture it.

30. Boyd, "Metaphor and Theory Change," 486. For an excellent discussion of constitutive metaphor, see Klamer and Leonard, "So What's an Economic Metaphor?"

31. Feedback metaphors differ from root metaphors *sensu stricto,* which were proposed by Pepper, in *World Hypotheses,* as the four fundamental and

largely incommensurable ways that we adopt to interpret the world: contextualism, formism, mechanism, and organicism. The metaphor of a root is itself a good one, as it provides an insight into how such large-scale metaphors draw ideas from afar into an established body of knowledge and, notwithstanding postmodernist arguments, provide an underlying meaning structure in modern society. Even more than literal roots, however, feedback metaphors tend to travel—they are nomadic by disposition and operate across the borders between usual domains, especially science and society. Furthermore, it is not clear that the metaphors considered here are as large-scale as root metaphors. Finally, Pepper's theory is limited by its Western and cognitivist bias, as explained by Buttimer, *Geography and the Human Spirit*, 84. For further reflection on how scholars such as Black and Pepper have thought about broad-scale metaphors, see Ricoeur, *Rule of Metaphor*, 244. On the significance of metaphors within Kuhnian paradigms, see Haraway, *Crystals, Fabrics, and Fields*. For further discussion of values in feedback metaphors, see chapter 3.

32. Fleming, "Can Nature (Legitimately) Be Our Guide?" Reflexivity is a critical component of the activism of a work such as mine; see, e.g., Woodhouse et al., "Science Studies and Activism."

33. Stepan, "Race and Gender," 271. This discussion demonstrates the proximity of the concepts of analogy and metaphor; Stepan did not distinguish between them in her paper.

34. The notion of self-fulfilling prophecies comes from Lakoff and Johnson, *Metaphors We Live By*, 156. That is, once a metaphor is chosen, social reality may more and more come to resemble it. For a start on the extensive discussion of the significance of evolutionary metaphors, see Greene, *Science, Ideology, and World View*; Young, *Darwin's Metaphor*; Moore, "Socializing Darwinism"; Keller, "Language and Ideology in Evolutionary Theory"; Richards, *Meaning of Evolution*; Taylor, "Natural Selection: A Heavy Hand"; Oyama, *Evolution's Eye*; Lewontin, *Biology as Ideology*; Goatly, *Washing the Brain*.

35. For a popular introduction to other cultures in this context, see Davis, *Wayfinders*. Note that scholars actively discuss and debate the notion of a gap between language and reality implied here; see, e.g., Szerszynski, "On Knowing What to Do"; Abram, *Spell of the Sensuous*. Postmodern cultural theorists would instead emphasize the nonreferential dimension of language and the ways that language and culture are "self-referential"; see, e.g., Livingston, *Between Science and Literature*.

36. For further discussion of biodiversity as a metaphor, see Väliverronen, "Biodiversity and the Power of Metaphor."

37. See Nordhaus and Shellenberger, *Break Through*.

II
Progress

1. Nisbet, *History of the Idea of Progress,* 4. The discussion of progress here draws from Bury, *Idea of Progress;* Wagar, *Good Tidings;* Almond, Chodorow, and Pearce, *Progress and Its Discontents;* Marx and Mazlish, *Progress.* My emphasis on Darwin is not meant to deny the existence of extensive thought much earlier, as far back as classical times, about the connections between progress in the natural and social worlds.

2. Ruse, *Monad to Man,* 526. This connection holds regardless of Darwin's equivocation about progress. The major conundrum with the linkage of biological and cultural progress is whether we are a simple extension of other forms of life on the planet, or whether our mental capacities and culture set us apart; see, e.g., Campbell, "Biological and Social Evolution"; Fracchia and Lewontin, "Does Culture Evolve?"

3. On eugenics, see Kevles, "Eugenics"; Paul, "Darwin, Social Darwinism and Eugenics." And see Williams, *Adaptation and Natural Selection.*

4. One reviewer asked why a "constructivist" argument would rely on a survey; the answer is that it is one way of finding out about the world, and in this case it served to provide a big-picture view of the status of these metaphors. The survey followed recommended protocols, including ethics approval, pretesting, and a design that sought to minimize respondent burden and thus maximize response rate; see, e.g., Schonlau, Fricker, and Elliot, *Conducting Research Surveys;* Sills and Song, "Innovations in Survey Research." Respondents were contacted in November 2003 with a personalized introductory e-mail that provided a link to one of two randomized versions of the survey, as well as a letter of endorsement from the president of their society. After two weeks, nonrespondents were sent a reminder e-mail. For further details on the survey, including discussion of its limitations, see Larson, "Social Resonance."

5. Cronbach's alpha, a measure of the extent to which survey statements cohere as a single factor, for the seven statements about progress, was an acceptable 0.76.

6. In the discussion below, note that "evolutionary biologists" and "evolutionaries" are shorthand for "respondents from the SSE" and "respondents from the FCE," respectively. The percentages reported here do not include respondents who chose the off-scale options provided in the survey questionnaire. Note that the sample size and response rate for the SSE were much greater than those for the FCE (789 versus 41 respondents, equivalent to respective response rates of 33.4 percent and 13.1 percent). These response rates are within the range reported for previous Web and e-mail surveys, but the

nonrespondents could introduce a systemic bias that must be kept in mind. Thus, I compare the organizations mostly impressionistically rather than with statistical tests. This quantitative study also limits my ability to interpret why people responded to statements the way that they did.

7. Membership count was at the time of the survey, at which point about 80 percent of its members were North American; www.evolutionsociety.org.

8. Of respondents from the SSE, 75.9 percent strongly agreed, compared to 42.5 percent from the FCE. Members of the SSE were much less likely to strongly agree with related statements, such as "Scientific knowledge gradually converges on the truth," with which 24.6 percent strongly agreed, and "Science will eventually answer all important questions about the universe, including humans," with which only 4.7 percent strongly agreed. Thus, they did not generally harbor a strong form of scientism; see Pigliucci, *Denying Evolution.*

9. Richards, "Evolution," 105; McShea, "Metazoan Complexity and Evolution," 488. For discussion of whether "complexity" is a metaphor, see Proctor and Larson, "Ecology, Complexity, and Metaphor." For further discussion of the issues about evolutionary complexity raised here, see Raup and Sepkowski, "Mass Extinctions"; Maynard Smith, "Evolutionary Progress"; Gould, "On Replacing the Idea of Progress"; Knoll and Bambach, "Directionality in the History of Life."

10. Of respondents from the FCE, 51.2 percent strongly agreed with the statement "I believe in God," compared to 5.9 percent from the SSE; 67.4 percent versus 6.3 percent of them strongly agreed with the statement "I would generally describe myself as spiritual." There is methodological risk here of reifying the distinction between a spiritual and a scientific group, but that is not my intention. Even among evolutionary biologists there is a historical association between biological progress and religion, which has much to do with its importance within evolutionary theory for so long, as well as its recent rejection. Julian Huxley and Theodosius Dobzhansky, for example, found support for their progressivism in religion, in contrast to Williams's somewhat Manichean motivation against progress; see, e.g., Ruse, "Molecules to Men."

11. "What Is Conscious Evolution," www.evolve.org/pub/doc/evolve _what_is_ce.html, quoting partly from Banathy, *Guided Evolution of Society;* This material was no longer available in August 2010, but similar material can be found at www.barbaramarxhubbard.com/con/node/8. The FCE had about three hundred members at the time of the survey. Their worldview is further explained by Hubbard, *Conscious Evolution.* Teilhard de Chardin, *Phenomenon of Man,* proposed an "Omega Point" toward which evolutionary processes aim. This proposal was met by vehement scientific critiques; see,

e.g., Medawar, "Review of the Phenomenon of Man"; Williams, *Adaptation and Natural Selection;* Monod, *Chance and Necessity.* Unsurprisingly, his system was also rejected by the Church. On the other visionaries mentioned, see Zaehner, *Evolution in Religion;* Ruse, *Monad to Man.*

12. Of FCE respondents, 30.8 percent strongly agreed with the first statement, compared to 13.0 percent from the SSE; 8.3 percent versus 1.0 percent of them strongly agreed with the second statement. On spiritual progress, see Menaker and Menaker, *Ego in Evolution;* Maslow, *Toward a Psychology of Being;* Fowler, *Stages of Faith;* Csikszentmihalyi, *Evolving Self.* For critique, see Rosenthal, *Words and Values.*

13. Chaisson, *Cosmic Evolution,* 3. For explication of these narratives, see Swimme and Berry, *Universe Story;* Capra, *Web of Life;* Earley, *Transforming Human Culture;* Genet, "Epic of Evolution"; Chaisson, "Ethical Evolution"; Rue, *Everybody's Story;* and for critical discussion see Toulmin, *Return to Cosmology;* Csikszentmihalyi, "Mythic Potential of Evolution." Universal progress is in part a derivative of the nebular hypothesis of the eighteenth century.

14. Genetic drift is a random process whereby gene frequencies in a population may change solely because of sampling effects. Of FCE respondents, 45.9 percent disagreed or strongly disagreed with this statement, compared to 26.6 percent from the SSE. 30.8 percent of respondents from the FCE strongly agreed with the statement "Evolutionary change requires intelligent design," compared to 1.2 percent from the SSE. And 36.6 percent versus 0.4 percent of them strongly agreed with the statement "Evolution has an aim or purpose." The extent to which they would discriminate between "design" as a general issue and its particular instantiation as a new-fangled creationist movement, intelligent design, was not addressed by the survey.

15. The etymology of metaphor is complex, and the meaning here is a simplified adaptation from *Webster's Third New International Dictionary.* Metaphor is also linked to the Greek terms *epiphora* and *phora,* which highlight the "'transference' or 'locomotion' of meanings across terms"; Baake, *Metaphor and Knowledge,* 62. For more on the view of metaphor presented here, see Bono, "Science, Discourse, and Literature"; Maasen and Weingart, "Metaphors—Messengers of Meaning"; Maasen, Mendelsohn, and Weingart, *Biology as Society, Society as Biology.*

16. See, e.g., Maasen and Weingart, "Metaphors—Messengers of Meaning"; Rozzi, "Reciprocal Links."

17. Williams, *Keywords,* 244.

18. I encountered the metaphor of adaptive radiation in this context in Collins and Kephart, "Science as News," 39. On evolutionary ecology, see note 17 in chapter 1.

19. Harrington, "Metaphoric Connections," 359–360. Her use of the

term *literal* reveals how easily even critics can be taken in by their metaphors. On the benefits of thinking of a web of life, see Marshall, *Nature's Web*.

20. See, e.g., Weingart and Maasen, "Order of Meaning." For a critique of systems thinking, see Berman, "Shadow Side of Systems Theory." For an introduction to the many applications of network thinking in ecology and evolution, see Proulx, Promislow, and Phillips, "Network Thinking in Ecology."

21. Star and Griesemer, "Institutional Ecology, 'Translations' and Boundary Objects," 393. On metaphors as boundary objects, see Väliverronen, "Biodiversity and the Power of Metaphor."

22. For the Templeton Foundation, see www.templeton.org.

23. Chaisson, *Cosmic Evolution*, 11. See also Kauffman, *Origins of Order*.

24. See, e.g., www.millenniumassessment.org/en/index.aspx.

25. Gibson, *Sustainability Assessment*, 46.

26. Nisbet, *History of the Idea of Progress*, 8.

27. Shanahan, "Evolutionary Progress?" 452.

28. Ruse, "Evolution and Progress."

29. For further discussion of this imagery, see Gould, *Wonderful Life;* Bowler, *Life's Splendid Drama;* Beer, *Darwin's Plots.*

30. See note 11 to this chapter. Also see Csikszentmihalyi, *Evolving Self;* Earley, *Transforming Human Culture;* Hubbard, *Conscious Evolution;* Banathy, *Guided Evolution of Society;* Csikszentmihalyi, "Mythic Potential of Evolution."

31. Gould, *Ever Since Darwin*, 37–38. Of FCE respondents, 32.4 percent agreed or strongly agreed with this statement compared to 4.6 percent from the SSE. Note, however, that 49 percent of them disagreed or strongly disagreed with it.

32. Pigliucci, *Denying Evolution*, 119, 130.

33. Kuhn, *Structure of Scientific Revolutions*, 170–173.

34. On the disunity of sciences, see Dupré, *Disorder of Things;* Feyerabend, "Quantum Theory and Our View of the World"; Galison and Stump, *Disunity of Science;* Cartwright, *Dappled World*. On the risk of applying science beyond its limits, see Berry, *Life Is a Miracle*. On career risks of interdisciplinarity, see Rhoten and Parker, "Risks and Rewards."

35. See www.icpsr.umich.edu.

36. Lakoff and Johnson, *Philosophy in the Flesh*, 63. The discussion here follows theirs, which may be consulted for further details.

37. Von Bertalanffy, "Essay on the Relativity of Categories," 258; Daston, "How Nature Became the Other," 39–40.

38. Hodge, "Natural Selection," 219. On personification in alternative conceptions of genes, see Hayles, "Desiring Agency." On personification in natural selection, see Young, "Darwin's Metaphor"; Taylor, "Natural Selection:

A Heavy Hand"; Oyama, *Evolution's Eye.* On Darwin's writing, see Young, "Darwin's Metaphor and the Philosophy of Science." For further discussion by science studies scholars of personification in a variety of biological fields, see Young, "Darwinism *Is* Social"; Midgley, *Science as Salvation;* Schiebinger, *Nature's Body;* Daston, "How Nature Became the Other."

39. Pigliucci, "Design Yes, Intelligent No," 38; Cassirer quoted in Ho and Fox, *Evolutionary Processes and Metaphors,* 3. On consciousness, see Schumacher, *Guide for the Perplexed;* Wallace, *Taboo of Subjectivity.*

40. Wright, "Panpsychism and Science," 82. On Darwin's panspychism, see Smith, "Charles Darwin."

41. Of FCE respondents, 57.5 percent strongly agreed with this statement, compared to 7.3 percent from the SSE.

42. De Waal, "Reading Nature's Tea Leaves," 66, 70. On Japanese primatology, see also Asquith, "Japanese Science and Western Hegemonies." On evolution and our relation with other organisms, see Singer, *Darwinian Left.*

43. Ingold, *Perception of the Environment,* 91, 50.

44. Trombulak, "Misunderstanding Neo-Darwinism," 1203; Feyerabend, *Conquest of Abundance,* 5–8, 145; Evernden, "Beyond Ecology," 101. See also Smith, "Metaphorical Basis of Selfhood."

45. On the tension between constitutive mechanist and organicist metaphors in the history of ecology, see Mitman, "From the Population to Society"; Taylor, "Technocratic Optimism"; Boucher, "Newtonian Ecology and Beyond"; Ulanowicz, "Life after Newton."

46. Harman and Sahtouris, *Biology Revisioned,* 12. On the death of nature, see Merchant, *Death of Nature.* Also see Abram, "Mechanical and the Organic"; Berman, *Reenchantment of the World.*

47. On machine people, see Harrington, "Metaphoric Connections." And see Taylor, "Technocratic Optimism."

48. Einarsson, "All Animals Are Equal but Some Are Cetaceans."

49. Midgley, *Science as Salvation,* 9, 67.

III

Competitive Facts and Capitalist Values

1. Weingart, "'Struggle for Existence,'" 130. On the view of competition presented here, also see Young, *Darwin's Metaphor: Nature's Place in Victorian Culture;* Young, "Darwinism *Is* Social"; McIntosh, "Competition." Note, however, that the romanticist idea of nature as harmonious was by no means uniform; see, e.g., Merchant, *Death of Nature.*

2. Hodge, "Natural Selection," 217; see also Paul, "Fitness." On the rela-

tion between natural selection and struggle, see Keller, "Competition: Current Usages"; Taylor, "Natural Selection: A Heavy Hand"; Rozzi, "Reciprocal Links"; Allchin, "More Fitting Analogy." The links between Darwinism per se and social Darwinism are complex and certainly not directly causal. Many historians have pointed out that Darwinism could be (and was) interpreted as support for an array of social policies, not just competitive and militaristic ones. For a discussion, see Rogers, "Darwinism and Social Darwinism"; Crook, *Darwinism, War and History;* Bowler, "Social Metaphors in Evolutionary Biology"; Weingart, "'Struggle for Existence'"; Caudill, *Darwinian Myths;* Hawkins, *Social Darwinism;* Paul, "Darwin, Social Darwinism and Eugenics." Some historians have even argued that Darwin himself was a social Darwinist; see, e.g., Moore, "Socializing Darwinism." For more on the applicability of Darwinian theory to social thought, see, among many others, Rose, *Darwin's Spectre;* Singer, *Darwinian Left.*

3. See, e.g., Merchant, "Radical Ecology." On the link between progress and competition, also see Goatly, *Washing the Brain.*

4. Some scholars would consider the derivation of what "ought to be" from what "is" an instance of what G. C. Moore dubbed the naturalistic fallacy; see, e.g., Wilson, Deitrich, and Clark, "On the Inappropriate Use of the Naturalistic Fallacy." Such concerns are more precisely attributable to Hume, however; the fallacy is applicable instead to the attempt to derive ethical properties from natural properties. Thus, I will not discuss the naturalistic fallacy here, even though it has been discussed extensively in the context of evolutionary thought; see, e.g., Farber, *Temptations of Evolutionary Ethics;* Maienschein and Ruse, *Biology and the Foundation of Ethics;* Rosenberg, *Darwinism in Philosophy.* For further discussion of the relation between is and ought in environmental metaphors, see Fleming, "Can Nature (Legitimately) Be Our Guide?"

5. Wilson, Deitrich, and Clark, "On the Inappropriate Use of the Naturalistic Fallacy," 671. On rape, see Zuk, "Feminism and the Study of Animal Behavior"; Dupré, "Fact and Value."

6. See, e.g., Solomon et al., *Climate Change 2007,* 5.

7. Dupré, "Fact and Value," 31. Though the distinction between facts and values is useful, its reification causes problems; see, e.g., Putnam, *Collapse of the Fact/Value Dichotomy;* Dupré, "Fact and Value." For further discussion of values in science, see below, and Shrader-Frechette and McCoy, "How the Tail Wags the Dog"; Proctor, "Expanding the Scope of Science and Ethics"; Kincaid, Dupré, and Wylie, "Value-Free Science?"; Douglas, *Science, Policy, and the Value-Free Ideal.* For climate change science specifically, see Demeritt, "Prospects for Constructivist Critique"; Pielke, *Honest Broker.*

8. For discussion and further examples, see Williams, *Ethics and the*

Limits of Philosophy; Norton, "Improving Ecological Communication"; Putnam, *Collapse of the Fact/Value Dichotomy;* Dupré, "Fact and Value."

9. Harré, Brockmeier, and Mühlhäusler, *Greenspeak,* 48; Schön, "Generative Metaphor," 150. For further discussion and examples of value resonance, see Bono, "Science, Discourse, and Literature"; Killingsworth and Palmer, *Ecospeak;* Baake, *Metaphor and Knowledge.* On value creep, see Carolan, "Values and Vulnerabilities of Metaphors." Technically, if a metaphor contains a normative element, it does not violate the distinction between is and ought because one is able to derive an ought from it alone. The key point here is to recognize how such metaphors entangle facts and values.

10. For discussion of the costs and benefits of the ecosystem health metaphor, see, e.g., Wicklum and Davies, "Ecosystem Health and Integrity?"; Ross et al., "Ecosystem Health Metaphor"; Lackey, "Values, Policy, and Ecosystem Health"; Davis and Slobodkin, "Science and Values of Restoration Ecology."

11. See, e.g., Richards, *Human Nature after Darwin.*

12. Carolan, "Society, Biology, and Ecology," 402.

13. "Human Behavior and Evolution Society," http://en.wikipedia.org/wiki/Human_Behavior_and_Evolution_Society (and formerly at www.hbes.com). Among the four organizations, 1,444 of 1,866 (77.4 percent) individuals who responded to this statement agreed or strongly agreed with it. The percentages for members of the specific organizations were SSE, 69.6; FCE, 73.8; HBES, 73.9; NABT, 86.6. The response rates for the HBES and NABT were 41.5 percent (317 respondents) and 23 percent (745 respondents), respectively. The responses to this statement are representative of the results for the other statements regarding competition in the survey, as detected by factor analysis. See note 4 in chapter 2 for details on the survey.

14. "Mission Statement," www.nabt.org/websites/institution/index.php?p=1; Schwartz, "Universals in the Context and Structure of Values," 18; www.icpsr.umich.edu. The GSS percentages were calculated for the 2,694 respondents who chose a position on the five-point scale. The remaining 22 percent chose "neither agree nor disagree." Clearly this statement is not directly comparable with the one in my survey.

15. Bunnell, "Attributing Nature with Justifications," 470. The respondents who disagreed or strongly disagreed with this statement represented 37.1 percent of those from the NABT, 55.1 percent from HBES, 56.8 percent from FCE, and 69.1 percent from the SSE. Leaving aside neutral responses, this represented a strong majority in each case except the NABT. I found a similar pattern for an array of additional statements containing competitive evolutionary metaphors. Respondents from each organization evaluated the potential social implications of nearly every statement less favorably than

its empirical basis, even though their consequences were not considered as negative as they were for struggle for survival.

16. A factor analysis united nine statements onto a "competition" factor. Cronbach's alpha for these nine statements was 0.64 (Q1) and 0.84 (Q2). Correlation between competition factor loadings for responses to Q1 and Q2: Spearman's *rho* = 0.42, $p < 0.001$, $n = 687$. I could use only an impressionistic ANCOVA to compare correlations among groups, since its assumptions were not met. Nonetheless, univariate correlations varied as follows: 0.30 (HBES), 0.35 (SSE), 0.41 (FCE), and 0.48 (NABT) (Spearman's *rho;* all $p < 0.001$ except FCE).

17. Only members of the FCE tended to agree or strongly agree with Q1, whereas members of the other groups more often disagreed or strongly disagreed (though only 45.0 percent of HBES members did so). Agreeing or strongly agreeing with Q2 were 44.0 percent of SSE members, 49.6 percent of HBES members, 55.1 percent of NABT members, and 84.4 percent of FCE members. Though the two statements were not semantic opposites, as this one concerned "animals" and the competitive one concerned "evolution," they still provide a general sense of the bias toward seeing the natural world in a competitive light.

18. Agreeing or strongly agreeing with the first and second statement, respectively, were 60 and 57.5 percent of FCE members, 80.1 and 73.7 percent of NABT members, 87.2 and 79.3 percent of HBES members, and 88.1 and 84.1 percent of SSE members.

19. Sober, "Kindness and Cruelty in Evolution," 59. On sperm competition, see Martin, "Egg and the Sperm." Note, however, that a more recent analysis of textbook presentations of the interaction between egg and sperm raises questions about the validity of Martin's analysis; P. J. Taylor, pers. comm., October 4, 2009.

20. Keller, "Demarcating Public from Private Values," 204; Singer, *Darwinian Left,* 19. On this discussion, also see Boucher, "Idea of Mutualism."

21. Keddy, *Competition,* 162–163.

22. Lakoff and Johnson, *Philosophy in the Flesh,* 558. On the competitive bias in our culture, see, among others, Caudill, *Darwinian Myths;* Taylor, "Natural Selection: A Heavy Hand"; Midgley, *Myths We Live By;* Allchin, "More Fitting Analogy." Most of this discussion follows Keller, "Demarcating Public from Private Values"; Keller, "Language and Ideology in Evolutionary Theory"; Keller, "Competition: Current Usages"; Oyama, *Evolution's Eye.* See also Evernden's discussion of *Umwelt* in *Natural Alien,* 146–148.

23. Boucher, "Idea of Mutualism," 22–23; Capra, *Web of Life,* 243. On competition as panchreston, see McIntosh, "Competition." On Newtonian ecology, see also Ulanowicz, "Life after Newton."

24. Sober, "Kindness and Cruelty in Evolution," 54.

25. Oyama, *Evolution's Eye*, 211; Margulis, "Words as Battle Cries," 673.

26. Rozzi et al., "Natural Drift." Also see Bunnell, "Attributing Nature with Justifications."

27. Dalai Lama, "Understanding Our Fundamental Nature," 78.

28. Proctor, *Value-Free Science?* 62.

29. Longino, "Gender and Racial Biases," 140–141. For Longino's full argument, see *Science as Social Knowledge*.

30. Midgley, *Myths We Live By,* 77; Putnam, *Collapse of the Fact/Value Dichotomy,* 32–33.

31. Proctor, *Value-Free Science?* 61

32. Lonergan, *Insight,* 545. On the relation among metaphor, model, and myth, see Livingstone and Harrison, "Meaning through Metaphor."

33. See the insightful discussion in Coyne, *Designing Information Technology*.

IV
Engaging the Metaphoric Web

1. Young, "Darwinism *Is* Social"; Young, "Darwin's Metaphor and the Philosophy of Science"; Harrington, "Science of Compassion"; Feyerabend, *Conquest of Abundance,* 247; Kitcher, *Science, Truth, and Democracy;* Kitcher, "Responsible Biology"; Bocking, *Nature's Experts,* 207. See also Douglas, "Moral Responsibilities of Scientists"; Douglas, *Science, Policy, and the Value-Free Ideal.*

2. For a review of critical discourse analysis, see Blommaert and Bulcaen, "Critical Discourse Analysis." On critical metaphor analysis, see Dirven, Hawkins, and Sandikcioglu, *Language and Ideology;* Dirven, Frank, and Ilie, *Language and Ideology.* For a classic example, see Chilton, *Security Metaphors.* Fill and Mühlhäusler's *Ecolinguistics Reader* provides an edited anthology for ecolinguistics and ecocritical linguistics.

3. Harré, Brockmeier, and Mühlhäusler, *Greenspeak,* 22. The discussion of words promoting exploitation follows Schultz, "Language and the Natural Environment."

4. Harré, Brockmeier, and Mühlhäusler, *Greenspeak,* 22.

5. Ibid.

6. Kinchy and Kleinman, "Organizing Credibility," 872. See also Takacs, *Idea of Biodiversity.*

7. Westoby, "What Does 'Ecology' Mean?"

8. Whitehead, *Science and the Modern World,* 181.

9. On the view of sustainability presented here, see Robinson, "Squaring the Circle?"; Kates, Parris, and Leiserowitz, "What Is Sustainable Development?"; Gibson, *Sustainability Assessment*. For an assessment of the term *sustainability* along the three axes of adequacy, see Penman, "Environmental Matters and Communication Challenges."

10. See, e.g., Walker and Salt, *Resilience Thinking;* Ehrenfeld, *Sustainability by Design.*

11. Bohm, *Wholeness and the Implicate Order.* For a technical discussion of whether English grammar, as represented in environmental science, is consistent with sustainability, see Goatly, "Green Grammar and Grammatical Metaphor"; Goatly, *Washing the Brain.*

12. Whorf, *Language, Thought, and Reality,* 213. Important recent collections on linguistic relativity include Gumperz and Levinson, *Rethinking Linguistic Relativity;* Gentner and Goldin-Meadow, *Language in Mind.*

13. Boroditsky, "Linguistic Relativity," 920. On Russian speakers, see Winawer et al., "Russian Blues Reveal Effects of Language."

14. Chawla, "Linguistic and Philosophical Roots," 254–255. Her paper suggests that these features of English are best shown through contrast with Amerindian languages, an argument fraught with generalization about these languages despite their considerable variation.

15. Ibid., 256. Also see Goatly, *Washing the Brain,* 324–325.

16. Goatly, *Washing the Brain,* 320, 315. Goatly's discussion of Niitsi'powahsin draws on Peat, *Blackfoot Physics.* Pertinent process philosophers include Henri Bergson, Charles Peirce, and Alfred North Whitehead. On Buddhism, see Macy, *Mutual Causality.*

17. Montgomery, "Of Towers, Walls, and Fields."

18. Harding, "Philosophies of Science."

19. Basso, *Wisdom Sits in Places,* 89–91.

20. Harrison, *When Languages Die,* 3, 17. For a critique of the metaphor of language "endangerment," see Hill, "'Expert Rhetorics' in Advocacy." The claim about Niitsi'powahsin comes from Goatly, *Washing the Brain,* 314.

21. For further discussion of how we might choose among metaphors, see Philippon, *Conserving Words,* 268–272. Also see the sage examination of metaphoric evaluation by Booth, "Metaphor as Rhetoric." On how economic metaphors suggest that Earth is a "welfare-producing machine," see Norton, "Beyond Positivist Ecology." On Earth as home, see Rowe, *Home Place.*

22. Santa Ana, *Brown Tide Rising,* 319; Blackmore, "Waking from the Meme Dream." On emancipatory discourse, see Janks and Ivanic, "Critical Language Awareness."

23. Underwood, "Toward a Poetics of Ecology"; Kolodny, *Lay of the Land.*

24. Torgerson, *Promise of Green Politics*, xi, 100.

25. This discussion is adapted from Grove-White, "Environmentalism: A New Moral Discourse." Also see Evernden, *Natural Alien*, 148–154.

26. Maguire, "Tears inside the Stone," 185. On the absence of management in the Swedish language, see Heberlein, "Wildlife Caretaking vs. Wildlife Management."

27. Norton and Noonan, "Ecology and Valuation," 673. See also Norton, *Sustainability;* Taylor, *Unruly Complexity;* Norton, "Beyond Positivist Ecology."

28. See, e.g., Boucher, "Newtonian Ecology and Beyond"; Bradshaw and Bekoff, "Ecology and Social Responsibility." There has been extensive debate about whether the loss of the idea of nature might undermine the justification for conservation. Thus, those who see humans as antithetical to nature would enforce the nature-culture duality; for discussion, see Jenkins, "Assessing Metaphors of Agency." For a critical realist theoretical framework that seeks to resolve this problem, see Carolan, "Society, Biology, and Ecology."

29. Ingold, *Perception of the Environment*, 42. For a review of nature-culture dualism in the context discussed here, see Frank, "Shifting Identities."

30. Bird-David, "Tribal Metaphorization." Also see essays in Ingold, *Perception of the Environment.*

31. Bertolas, "Cross-Cultural Environmental Perception," 100, 109. This discussion touches on the extensive debate about the social construction of nature; a classic paper on wilderness is Cronon, "Trouble with Wilderness." Also see responses such as Soulé and Lease, *Reinventing Nature;* Crist, "Against the Social Construction of Nature."

32. Chaisson, "Ethical Evolution," 271; Callicott, "Multicultural Environmental Ethics." On the cost of such large-scale ideologies, see Bellah, *Beyond Belief;* Toulmin, *Return to Cosmology.*

33. Dickinson, "Lyric Ethics," 39–42, drawing on Zwicky, *Wisdom & Metaphor,* a playful yet important exploration of how metaphor helps us understand the world.

34. Many scholars in science studies have attempted to subvert such dualities, but see in particular Haraway, *Modest_Witness@Second_Millennium.*

35. On partnership ethics, see Merchant, "Partnership Ethics and Cultural Discourse."

36. Evernden, "Beyond Ecology," 99, 95. On the microbiome, see http://nihroadmap.nih.gov/hmp.

37. Oliver, *Leaf and the Cloud,* 27. For a summary of the anthropological view, see Geertz, *Local Knowledge,* 59.

38. Hesse, "Explanatory Function of Metaphor," 259; Evernden, "Beyond Ecology," 102.

39. For further discussion of the global environment, see Berry, *Life Is a Miracle;* Proctor, "Environment after Nature," and Ingold, *Perception of the Environment,* chap. 12. On hot spots, see Carolan, "'This Is Not a Biodiversity Hotspot.'"

40. See, e.g., Buber, *I and Thou.*

41. Ehrenfeld, "Industrial Ecology"; Philippon, *Conserving Words.* For reflections on the suitability of the island metaphor for fragmentation, see Haila, "Conceptual Genealogy of Fragmentation." For discussion of chaos as a guiding metaphor, see Fleming, "Can Nature (Legitimately) Be Our Guide?"

42. Jenkins, "Assessing Metaphors of Agency." Also see Guha, "Authoritarian Biologist"; Robbins, "Tracking Invasive Land Covers in India."

43. Neumann, "Moral and Discursive Geographies."

44. On the two case studies here, see Webb and Raffaelli, "Conversations in Conservation"; Hardy-Short and Short, "Fire, Death, and Rebirth." "Rhetorical jujitsu" comes from Beardsley, "Framing Biology." For further discussion of language conflict and potential solutions, see Schön and Rein, *Frame Reflection;* Lewicki, Gray, and Elliott, *Making Sense of Intractable Environmental Conflicts.*

V
When Scientists Promote

1. Väliverronen, "Biodiversity and the Power of Metaphor"; Collins and Kephart, "Science as News." A significant challenge is that biodiversity can be defined in so many ways, even among ecologists; see, e.g., Holt, "Biodiversity Definitions Vary within the Discipline."

2. Yoon, *Naming Nature,* 281–282.

3. Edge, "Technological Metaphor and Social Control," 136.

4. See Buttimer, *Geography and the Human Spirit;* Fine and Sandstrom, "Ideology in Action"; Trudgill, "Psychobiogeography"; Nelkin, "Molecular Metaphors."

5. Fine and Sandstrom, "Ideology in Action," 27.

6. Barbour, "Ecological Fragmentation in the Fifties," 254; Fleck, *Genesis and Development of a Scientific Fact.* On Clements and Gleason, see Journet, "Ecological Theories as Cultural Narratives." On the hard core, as discussed by Imre Lakatos, see Klamer and Leonard, "So What's an Economic Metaphor?"

7. See, e.g., Yearley, *Sociology, Environmentalism, Globalization.*

8. Dawkins, *Selfish Gene;* Richards, *Human Nature after Darwin,* 165. In agreement or strong agreement with the statement were 12.8 percent of

FCE respondents, 25.1 percent from NABT, and 33.9 percent from SSE. Note, however, that this statement is double-barreled, since interpretation of the subordinate clause "because of their selfish genes" could depend on whether a respondent agreed that animals are competitive.

9. Bono, "Metaphorics of Scientific Practice," 227.

10. Nelkin, "Promotional Metaphors," 25.

11. Hebert et al., "Biological Identifications through DNA Barcodes." The citation count was 1035 according to Web of Science, October 31, 2010. On the aha! moment in a supermarket, see Stoeckle and Hebert, "Barcode of Life." I conducted a one-hour interview with Paul Hebert in January 2007, after receiving approval from the University of Waterloo Office of Research Ethics. I asked a number of open-ended questions, recorded and transcribed the responses, and then analyzed them for emergent themes. Unless indicated otherwise, quotes below are taken from this interview, though they may be slightly reworded in the interest of clarity (though without changing meaning).

12. Wheeler, "Losing the Plot," 406. Some of the critiques include Ebach and Holdrege, "More Taxonomy, Not DNA Barcoding"; Meyer and Paulay, "DNA Barcoding"; Rubinoff, "Utility of Mitochondrial DNA Barcodes"; de Carvalho et al., "Systematics Must Embrace Comparative Biology."

13. Takacs, *Idea of Biodiversity*, 8.

14. On the vision of a Life Barcoder, see Janzen, "Now Is the Time." On the cost of the project, see Hebert et al., "Biological Identifications through DNA Barcodes"; Hebert and Gregory, "Promise of DNA Barcoding."

15. See Rosner and Johnson, "Telling Stories."

16. On the tree of life, see McInerney, Cotton, and Pisani, "Prokaryotic Tree of Life."

17. See the document outlining the ten reasons for DNA barcoding at http://phe.rockefeller.edu/barcode. The two assumptions discussed here appear elsewhere in the literature on DNA barcoding, including Janzen, "Now Is the Time"; Hebert and Gregory, "Promise of DNA Barcoding." For further discussion, also see Larson, "DNA Barcoding"; Ellis, Waterton, and Wynne, "Twenty-first-Century DNA Barcoding."

18. "DNA 'Bar Codes' Identify 15 New Species of Birds," www.cbc.ca/technology/story/2007/02/19/science-dnabarcode.html. On maps, see Duncan, "Mapping Whose Reality?"

19. See Hebert et al., "Identification of Birds through DNA Barcodes."

20. Janzen, "Now Is the Time."

21. On grassroots initiatives, see Borgerhoff Mulder and Coppolillo, *Conservation*. For further review of "ecology-first" versus "people-included" approaches to conservation, see Stoll-Kleemann and O'Riordan, "From Par-

ticipation to Partnership in Biodiversity Protection"; Caillon and Degeorges, "Biodiversity." The arguments for earlier engagement are reviewed in Fiorino, "Citizen Participation and Environmental Risk."

22. Medley and Kalibo, "Global Localism," 159.

23. Nabhan and St. Antoine, "Loss of Floral and Faunal Story"; Basso, *Wisdom Sits in Places;* Henderson, "Ayukpachi."

24. Dauvergne, *Shadows of Consumption.*

25. For elaboration, see Callon, "Elements of a Sociology of Translation."

26. Bowker and Star, *Sorting Things Out,* 5–6. On molecular biology and life, see Kay, *Who Wrote the Book of Life?* For a similar argument for ecological systems theorists, see Taylor, "Technocratic Optimism." On the species concept in DNA barcoding, see Meyer and Paulay, "DNA Barcoding"; Rubinoff, "Utility of Mitochondrial DNA Barcodes"; Fitzhugh, "DNA Barcoding."

27. Winner, *Whale and the Reactor,* 99. On global technological oversight, see Edwards, *Closed World.* On taxonomic inflation, see Isaac, Mallett, and Mace, "Taxonomic Inflation."

28. Wilson, *Naturalist,* 11–12. See also Ebach and Holdrege, "More Taxonomy, Not DNA Barcoding"; Holloway, "Democratizing Taxonomy." Yoon, *Naming Nature,* goes even further, developing a persuasive argument that scientific taxonomy disconnects us from the natural world in general because its insights clash with our intuition for natural categories, her core example being that the group "fish" is scientifically invalid.

29. Pergams and Zaradic, "Love of Nature in the US," 387. On the need for naturalists, see Raven, "Taxonomy"; Miller, "Biodiversity Conservation."

30. The "reading" metaphor here is intriguing, not least because many cultures can read those species with which they interact quite effectively, though contemporary Western scientific culture is quite distinctive in its desire to capture everything within one grand, universalizing system. It coheres with the metaphor of biodiversity as a library containing information that only scientists can read; see, e.g., Väliverronen and Hellsten, "Metaphors in Communicating Biodiversity." For further perspective on this metaphor, see the critical discourse about reading the genome, e.g., Rosner and Johnson, "Telling Stories"; Kay, *Who Wrote the Book of Life?;* Dorries, *Experimenting in Tongues.*

31. Åkerman, "What Does 'Natural Capital' Do?" 438, 440. For discussion of the dominance of economic metaphors in ecology (e.g., producers and consumers), see Worster, *Nature's Economy.*

32. Norton, *Sustainability,* 311; Foster, "Making Sense of Stewardship," 32. For discussion of the benefits of thinking of ecosystem services, see Armsworth et al., "Ecosystem-Service Science."

33. Evans, "Biodiversity: Nature for Nerds?" 9.

34. Heidegger, *Question Concerning Technology*.

35. Sarewitz, *Frontiers of Illusion*, 142.

36. Quoted in Roberts, "Barcoding Life," 48.

37. Hebert and Barrett, "Reply to the Comment by L. Prendini," 505. On the printing press, see Janzen cited in Holloway, "Democratizing Taxonomy."

38. On constraints on scientific metaphors, see Toulmin, "Construal of Reality"; Hayles, "Desiring Agency." On hubris versus humility, see Jasanoff, "Technologies of Humility."

39. Young, "Darwin's Metaphor and the Philosophy of Science," 394. On puzzle solving in normal science, see Kuhn, *Structure of Scientific Revolutions*.

40. I thank Peter Taylor for this suggestion.

41. See Louv, *Last Child in the Woods*.

VI
Advocating with Fear

1. For a review, see Larson, "Alien Approach to Invasive Species."

2. On political geography, see Moore, "Revolution of the Space Invaders." On Nazi invasion, see Davis, Thompson, and Grime, "Dissociation of Invasion Ecology." On the conflation of spread with impact, see Ricciardi and Cohen, "Invasiveness of an Introduced Species." For an analysis of the conceptual metaphors of invasion biology, see Larson, "Biological, Cultural, and Linguistic Origins"; Larson, "Embodied Realism and Invasive Species."

3. Wittgenstein, *Philosophical Investigations*, 41; Otis, *Membranes*, 168. On tax cuts, see Lakoff, *Don't Think of an Elephant!* On boundaries in invasion biology, see Milton, "Ducks Out of Water"; Larson, "Embodied Realism and Invasive Species."

4. Simberloff, "Invasional Meltdown 6 Years Later," 912. The original paper was Simberloff and Von Holle, "Positive Interactions of Nonindigenous Species." Simberloff discussed the origin of this metaphor in our interview in Ottawa, Ontario, in February 2007. This interview had received approval from the University of Waterloo Office of Research Ethics. I asked a number of open-ended questions, recorded and transcribed the responses, and then analyzed them for emergent themes. Unless indicated otherwise, quotes below are taken from this interview, though they may be slightly reworded in the interest of clarity (though without changing meaning). On mutational meltdown, see Gabriel, Lynch, and Burger, "Muller's Ratchet and Mutational Meltdowns."

5. Simberloff, "Invasional Meltdown 6 Years Later," 916.

6. Ibid., 912; Gurevitch, "Commentary on Simberloff," 920.

7. For a review of problems with the dichotomy between native and introduced species, see Larson, "Who's Invading What?" But contrast with Simberloff, "Non-Native Species *Do* Threaten the Natural Environment!"

8. Simberloff made the remark about "a number of boring titles" during our interview; Simberloff, "Invasional Meltdown 6 Years Later," 916; Gurevitch, "Commentary on Simberloff," 919. And see Hardin, "Tragedy of the Commons"; Ehrlich and Ehrlich, *Extinction.*

9. Simberloff, "Invasional Meltdown 6 Years Later," 917. On the link between information and action, see, e.g., the introductory chapter in Moser and Dilling, *Creating a Climate for Change.* Also see the testimonial by a biologist, Allendorf, "Conservation Biologist as Zen Student."

10. Mio, "Metaphor and Politics," 128; Gobster, "Invasive Species as Ecological Threat," 264. Note, however, that there are methodological challenges with extrapolating psychological studies of fear appeals to real-world situations; for a review, see Hastings, Stead, and Webb, "Fear Appeals in Social Marketing."

11. O'Neill and Nicholson-Cole, "Fear Won't Do It," 375–376. Also see Moser and Dilling, *Creating a Climate for Change;* Nisbet, "Ethics of Framing Science." For a more radical perspective on fear than will be considered here, see Maguire, "Tears inside the Stone." In brief, he argues that widespread fear in our culture derives from the nature of our childhood (drawing on the work of the psychologist Alice Miller) and that such fear is on the one hand a method of social control and on the other a continual source of apathy. In both respects, it plays a critical role in our approach to environmental ills, not least through its connection with militarism.

12. Simberloff, "Invasional Meltdown 6 Years Later," 912; Pielke, *Honest Broker,* 134. On promotional metaphors, see Nelkin, "Promotional Metaphors."

13. B. Larson and C. Glass, unpublished data. Ninety percent agreed or strongly agreed with the statement, "I am concerned about the environmental impact of invasive alien species in Canada." The second statement was introduced as follows: "Scientists have coined the phrase 'invasional meltdown' to refer to 'the process by which non-indigenous species facilitate one another's invasion in various ways . . . potentially leading to an accelerating increase in number of introduced species and their impact.' This phrase has 'attracted great attention' despite the fact that 'a full "invasional meltdown" . . . has yet to be conclusively demonstrated.' Do you think it is appropriate for scientists to refer to this phenomenon as a 'meltdown?' Why or why not?" More than half, 51 percent, of the students disagreed or strongly disagreed, whereas only 23 percent agreed or strongly agreed with the use of this phrase. The remaining 26 percent were neutral (choosing 3 on the 5-point scale). The students

were not taught about invasional meltdown and hence had no particular implicit bias.

14. For further details see A. Young and B. Larson, in preparation. The survey was conducted using SurveyMonkey.com in April 2008. After pretesting, it was sent to the distribution list of reviewers of the journal *Biological Invasions*. Respondents were asked to evaluate these and other statements taken from refereed papers. There were 422 responses, giving a very good response rate of 42.5 percent. Ninety-seven percent of respondents either self-identified as invasion biologists or studied invasive species. Of those responding, 26.6 percent agreed or strongly agreed with the first statement here, whereas 48.3 percent disagreed or strongly disagreed. For the second statement, the respective percentages were 61.2 percent and 21.5 percent. The remaining respondents chose the neutral option (3 on the 5-point scale).

15. Hastings, Stead, and Webb, "Fear Appeals in Social Marketing," 973. On children's fear of nature, see Bixler and Floyd, "Nature Is Scary"; Louv, *Last Child in the Woods*.

16. Allan, "Exploring Natural Resource Management," 360.

17. Moser and Dilling, *Creating a Climate for Change*, 15.

18. Simberloff, "Invasional Meltdown 6 Years Later," 912.

19. Ibid., 916.

20. For a review and examples, see, among others, Chew and Laubichler, "Natural Enemies"; Larson, "War of the Roses"; Clergeau and Nuñez, "Language of Fighting Invasive Species."

21. Ornstein and Ehrlich, *New World, New Mind*.

22. See Glotfelty, "Cold War, Silent Spring."

23. Baskin, *Plague of Rats and Rubbervines*, 13.

24. A later version of the paper was published as Larson, "War of the Roses."

25. For analogous arguments against the use of militaristic metaphors in biomedicine, see Ross, "Militarization of Disease"; Sontag, *Illness as Metaphor*; Montgomery, *Scientific Voice*; McMichael, "Fine Battlefield Reporting."

26. Haila, "Beyond the Nature-Culture Dualism," 158. See also Haila, "Biodiversity and the Divide between Culture and Nature." On nature-culture hybridity, see Latour, *We Have Never Been Modern*.

27. For a review of the points mentioned here, see Larson, "Who's Invading What?" Some seminal early papers along these lines include MacDougall and Turkington, "Are Invasive Species the Drivers?"; Meiners, "Native and Exotic Plant Species." On positive roles of invasive species, see Ewel and Putz, "Place for Alien Species"; Pawson et al., "Non-Native Plantation Forests."

28. Muirhead et al., "Modelling Local and Long-Distance Dispersal," 76, emphasis added. On the continuum of disturbance, see Hull and Rob-

ertson, "Language of Nature Matters"; Brunson, "Managing Naturalness as a Continuum"; Ridder, "Value of Naturalness and Wild Nature."

29. Glover, "War on _____"; Dedaic and Nelson, *At War with Words*.

30. Jane Braxton Little, "What to Do about a Nasty Fish," *High Country News*, June 23, 1997, www.hcn.org/issues/110/3468; Jenifer Ragland, "U.S. Plan to Kill Rats on Anacapa Delayed a Month," *Los Angeles Times*, November 6, 2001, http://articles.latimes.com/2001/nov/06/local/me-839; Mackenzie and Larson, "Participation under Time Constraints."

31. Pfeiffer and Voeks, "Biological Invasions and Biocultural Diversity," 281. On views of nature, see Macnaughten and Urry, *Contested Natures*. For the Eastern Cape example, see Shackleton et al., "Assessing the Effects of Invasive Alien Species"; also see Hall, "Cultural Disturbances and Local Ecological Knowledge." I do not have space here to review the many recent papers emphasizing social dimensions of invasive species control; see Maguire, "Invasive Species Management"; Foster and Sandberg, "Friends or Foe?"; Robbins, "Comparing Invasive Networks"; Stokes et al., "Importance of Stakeholder Engagement"; Norgaard, "Politics of Invasive Weed Management"; Fischer and van der Wal, "Invasive Plant Suppresses Charismatic Seabird"; Evans, Wilkie, and Burkhardt, "Adaptive Management of Nonnative Species."

32. O'Brien, "Exotic Invasions, Nativism, and Ecological Restoration," 69. For a review of charges of xenophobia against invasion biology, see Simberloff, "Confronting Invasive Species." For further discussion of the connection between its metaphors and anti-immigrant rhetoric, see Fine and Christoforides, "Dirty Birds, Filthy Immigrants"; Pauly, "Beauty and Menace of the Japanese Cherry Trees"; Coates, *American Perceptions*. On metaphors for immigrants to the United States, see Santa Ana, "'Like an Animal I Was Treated.'"

33. Nelson, "Conclusion," 449; Glotfelty, "Cold War, Silent Spring," 168. On the "switch," see Underhill, "Metaphorical Representation of the War in Iraq." On connections between wars on drugs, poverty, and others, see Glover, "War on _____." On the war-insecticide link, see Russell, "'Speaking of Annihilation.'" On the environmental cost of war, see Austin and Bruch, *Environmental Consequences of War*.

34. Bardwell, "Problem-Framing."

35. For further thoughts on how to challenge metaphors, see Chilton, *Security Metaphors*, 70–71.

36. Lakoff, *Don't Think of an Elephant!*

37. Colautti and Richardson, "Subjectivity and Flexibility in Invasion Terminology," 1228. See the proposal for a neutral language by Colautti and MacIsaac, "Neutral Terminology." For a review, see Larson, "Alien Approach to Invasive Species."

38. Hodges, "Defining the Problem."

39. Colautti and Richardson, "Subjectivity and Flexibility in Invasion Terminology," 1228; van Fraassen, "World of Empiricism," 130. On the politics of definition, see Schiappa, "Towards a Pragmatic Approach to Definition."

40. Bradshaw and Bekoff, "Ecology and Social Responsibility," 462. On values in the crisis discipline of conservation biology, see Barry and Oelschlaeger, "Science for Survival."

41. Avise, "Evolving Genomic Metaphors," 86. For further discussion of alternative metaphors, also see Gobster, "Invasive Species as Ecological Threat"; Larson, "War of the Roses"; Larson, "Thirteen Ways of Looking at Invasive Species"; Keulartz and van der Weele, "Framing and Reframing in Invasion." On human security metaphors, see Tanentzap et al., "Human Security Framework." On immune system metaphors, see Walker, "Ecosystems and Immune Systems." On driver-and-passenger metaphors, see MacDougall and Turkington, "Are Invasive Species the Drivers?" On pressure-and-resistance metaphors, see Larson, "Embodied Realism and Invasive Species."

42. Propagule pressure can be defined as the quality, quantity, and frequency of invading organisms; biotic resistance describes the ability of resident species to prevent these newcomers from entering a community. On compositionalism and functionalism, see Callicott, Crowder, and Mumford, "Current Normative Concepts in Conservation." For an example of function maintained in a nonnative community, see Wilkinson, "The Parable of Green Mountain."

43. Schlaepfer et al., "Introduced Species as Evolutionary Traps."

44. Hulme, "Conquering of Climate," 12–14.

VII
Seeking Sustainable Metaphors

1. Condit, "Meaning and Effects of Discourse about Genetics." On the challenge of reflexive attention to one's values, see Blommaert and Bulcaen, "Critical Discourse Analysis."

2. See, e.g., Eden, "Public Participation in Environmental Policy"; Bocking, Nature's Experts.

3. This discussion is drawn from Collins and Evans, "Third Wave of Science Studies." Also see Demeritt, "Prospects for Constructivist Critique."

4. Sarewitz, "How Science Makes Environmental Controversies Worse."

5. For review of deliberative methods, see Douglas, "Inserting the Public into Science"; Einsiedel, "Public Engagement and Dialogue."

6. See also Elliott, "Ethical Significance of Language."

7. Nisbet, "Ethics of Framing Science," 57.

8. See Douglas, "Moral Responsibilities of Scientists"; Douglas, "Inserting the Public into Science."

9. See, e.g., Kitcher, *Science, Truth, and Democracy*. And see chapter 3 in this volume.

10. On the idea of extended peer review, see Funtowicz and Ravetz, "Science for the Postnormal Age."

11. Novacek, "Engaging the Public in Biodiversity," 11572.

12. See, e.g., Carolan, "Democratization of the Decision-Making Process." Such sociological analysis can complement the philosophical work recommended by Elliott, "Ethical Significance of Language."

13. On social marketing, see McKenzie-Mohr and Smith, *Fostering Sustainable Behavior*. The National Academies study is reviewed in Nisbet, "Ethics of Framing Science."

14. Gurevitch, "Commentary on Simberloff," 919.

15. Whitmarsh, "What's in a Name?"

16. Condit, "Meaning and Effects of Discourse about Genetics," 392.

17. Hardin, "Tragedy of the Commons"; Mio, Thompson, and Givens, "Commons Dilemma as Metaphor."

18. Schultz and Zelezny, "Reframing Environmental Messages."

19. Cilliers, *Complexity and Postmodernism*.

20. Zimmerman and Cuddington, "Students' Definitions of the 'Balance of Nature,'" 393. On scientists' concerns about lay consumption, see Young and Matthews, "Experts' Understanding of the Public."

21. López, "Notes on Metaphors, Notes as Metaphors," 7, 21; Hodges, "Defining the Problem," 39.

22. Condit et al., "Lay People's Understanding," 249.

23. Foster, "Making Sense of Stewardship," 32.

24. See, e.g., Carew and Mitchell, "Metaphors Used by Some Engineering Academics."

25. See Abram, "Mechanical and the Organic."

26. Fairclough, *Critical Language Awareness*, 6.

27. Frost, "Education by Poetry," 106–107. On language as a living being, see Livingston, *Between Science and Literature*.

28. Kuhn, *Structure of Scientific Revolutions*, 37. For an insightful discussion of the contingency of the language used in science, see Rorty, *Contingency, Irony, and Solidarity*.

29. Allison, Beggan, and Midgley, "Quest for 'Similar Instances,'" 484.

30. Minteer, Collins, and Bird, "Editors' Overview," 479. See also van Houtan, "Conservation as Virtue."

31. On language awareness in education, see Fairclough, *Critical*

Language Awareness. On scientific language, see, e.g., Harré, Brockmeier, and Mühlhäusler, *Greenspeak;* Fill and Mühlhäusler, *Ecolinguistics Reader;* Goatly, *Washing the Brain.*

32. I thank Cor van der Weele for suggesting this analogy.

33. See, e.g., Tobin and Tippins, "Metaphors as Seeds." For one model of the use of analogy in biology education, see Venville and Treagust, "Analogies in Biological Education." And see Cobern, "Science Education."

34. Bowers, "How Language Limits Our Understanding," 144–145. See also Armstrong, "Social Metaphors."

35. See note 28 in chapter 1.

36. Brand and Jax, "Focusing the Meaning(s) of Resilience." See also Carpenter et al., "From Metaphor to Measurement"; Pickett, Cadenasso, and Grove, "Resilient Cities."

VIII
Wisdom and Metaphor

1. See Osowski, "Ensembles of Metaphor"; Allison, Beggan, and Midgley, "Quest for 'Similar Instances.'"

2. Schön, "Generative Metaphor," 151. See also Schön and Rein, *Frame Reflection.*

3. Cook, *Genetically Modified Language,* 114.

4. Berry, *Life Is a Miracle,* 148. On Whitman's poem, see Maguire, "Tears inside the Stone." On the new vision provided by poetic metaphors, see Rorty, *Contingency, Irony, and Solidarity.*

5. Among others on this point, see the poetic pieces by Evernden, *Natural Alien;* Abram, *Spell of the Sensuous;* Fleischner, "Natural History."

Bibliography

Abram, D. "The Mechanical and the Organic: On the Impact of Metaphor in Science." In *Scientists on Gaia*, edited by S. H. Schneider and P. J. Boston. Cambridge, MA: MIT Press, 1993.

———. *The Spell of the Sensuous: Perception and Language in a More-Than-Human World*. New York: Vintage Books, 1997.

Åkerman, M. "What Does 'Natural Capital' Do? The Role of Metaphor in Economic Understanding of the Environment." *Environmental Values* 12:431–448, 2003.

Allan, C. "Exploring Natural Resource Management with Metaphor Analysis." *Society & Natural Resources* 20:351–362, 2007.

Allchin, D. "A More Fitting Analogy: How Does One Aptly Characterize Natural Selection?" *American Biology Teacher* 69:174–176, 2007.

Allendorf, F. W. "The Conservation Biologist as Zen Student." *Conservation Biology* 11:1045–1046, 1997.

Allison, S. T., J. K. Beggan, and E. H. Midgley. "The Quest for 'Similar Instances' and 'Simultaneous Possibilities': Metaphors in Social Dilemma Research." *Journal of Personality and Social Psychology* 71:479–497, 1996.

Almond, G. A., M. Chodorow, and R. H. Pearce, eds. *Progress and Its Discontents*. Berkeley and Los Angeles: University of California Press, 1982.

Armstrong, C. "Social Metaphors and Their Implications for Environmental Education." *Environmental Education Research* 3:29–42, 1997.

Armsworth, P. R., K. M. A. Chan, G. C. Daily, P. R. Ehrlich, C. Kremen, T. H. Ricketts, and M. A. Sanjayan. "Ecosystem-Service Science and the Way Forward for Conservation." *Conservation Biology* 21:1383–1384, 2007.

Asquith, P. J. "Japanese Science and Western Hegemonies: Primatology and the Limits Set to Questions." In *Naked Science: Anthropological Inquiry into Boundaries, Power, and Knowledge,* edited by L. Nader. New York: Routledge, 1996.

Austin, J. E., and C. E. Bruch, eds. *The Environmental Consequences of War: Legal, Economic and Scientific Perspectives.* Cambridge: Cambridge University Press, 2000.

Austin, J. L. *How to Do Things with Words.* Oxford: Clarendon, 1962.

Avise, J. C. "Evolving Genomic Metaphors: A New Look at the Language of DNA." *Science* 294:86–87, 2001.

Baake, K. *Metaphor and Knowledge: The Challenges of Writing Science.* Albany: State University of New York Press, 2003.

Balmford, A., and W. Bond. "Trends in the State of Nature and Their Implications for Human Well-Being." *Ecology Letters* 8:1218–1234, 2005.

Banathy, B. H. *Guided Evolution of Society: A Systems View.* Dordrecht: Kluwer, 2000.

Barbour, M. G. "Ecological Fragmentation in the Fifties." In *Uncommon Ground: Rethinking the Human Place in Nature,* edited by W. Cronon. New York: W. W. Norton, 1995.

Barcelona, A., ed. *Metaphor and Metonymy at the Crossroads: A Cognitive Perspective.* New York: Mouton de Gruyter, 2000.

Bardwell, L. V. "Problem-Framing: A Perspective on Environmental Problem-Solving." *Environmental Management* 15:603–612, 1991.

Barry, D., and M. Oelschlaeger. "A Science for Survival: Values and Conservation Biology." *Conservation Biology* 10:905–911, 1996.

Baskin, Y. *A Plague of Rats and Rubbervines: The Growing Threat of Species Invasions.* Washington, DC: Island Press, 2002.

Basso, K. H. *Wisdom Sits in Places: Landscape and Language among the Western Apache.* Albuquerque: University of New Mexico Press, 1996.

Bazerman, C. *Shaping Written Knowledge: The Genre and the Activity of the Experimental Article in Science.* Madison: University of Wisconsin Press, 1988.

Bazzaz, F., G. Ceballos, M. Davis, R. Dirzo, P. R. Ehrlich, T. Eisner, S. Levin, et al. "Ecological Science and the Human Predicament." *Science* 282:879, 1998.

Beardsley, T. M. "Framing Biology." *BioScience* 56:555, 2006.

Beck, U. "From Industrial Society to the Risk Society: Questions of Survival, Social Structure, and Ecological Enlightenment." *Theory, Culture & Society* 9:97–123, 1992.

Beer, G. *Darwin's Plots: Evolutionary Narrative in Darwin, George Eliot and Nineteenth-Century Fiction,* 2nd ed. Cambridge: Cambridge University Press, 2000.

Bellah, R. N. *Beyond Belief: Essays on Religion in a Post-Traditional World.* New York: Harper & Row, 1970.

Berman, M. *The Reenchantment of the World.* Ithaca, NY: Cornell University Press, 1981.

———. "The Shadow Side of Systems Theory." *Journal of Humanistic Psychology* 36:28–54, 1996.

Berry, W. *Life Is a Miracle: An Essay against Modern Superstition.* Washington, DC: Counterpoint, 2000.

Bertolas, R. J. "Cross-Cultural Environmental Perception of Wilderness." *Professional Geographer* 50:98–111, 1998.

Bird-David, N. "Tribal Metaphorization of Human-Nature Relatedness: A Comparative Analysis." In *Environmentalism: The View from Anthropology,* edited by K. Milton. New York: Routledge, 1993.

Bixler, R. D., and M. F. Floyd. "Nature Is Scary, Disgusting, and Uncomfortable." *Environment and Behavior* 29:443–467, 1997.

Black, M. *Models and Metaphors.* Ithaca, NY: Cornell University Press, 1962.

Blackmore, S. "Waking from the Meme Dream." In *The Psychology of Awakening: Buddhism, Science, and Our Day-to-Day Lives,* edited by G. Watson, S. Batchelor, and G. Claxton. York Beach, ME: Samuel Weiser, 2000.

Blommaert, J., and C. Bulcaen. "Critical Discourse Analysis." *Annual Review of Anthropology* 29:447–466, 2000.

Bocking, S. *Nature's Experts: Science, Politics, and the Environment.* New Brunswick, NJ: Rutgers University Press, 2004.

Bohm, D. *Wholeness and the Implicate Order.* New York: Routledge, 1980.

Bono, J. J. "Science, Discourse, and Literature: The Role/Rule of Metaphor in Science." In *Literature and Science: Theory and Practice,* edited by S. Peterfreund. Boston: Northeastern University Press, 1990.

———. "Why Metaphor? Toward a Metaphorics of Scientific Practice." In *Science Studies: Probing the Dynamics of Scientific Knowledge,* edited by S. Maasen and M. Winterhager. Bielefeld: Transcript Verlag, 2003.

Booth, W. C. "Metaphor as Rhetoric: The Problem of Evaluation." In *On Metaphor,* edited by S. Sacks. Chicago: University of Chicago Press, 1978.

Borgerhoff Mulder, M., and P. Coppolillo. *Conservation: Linking Ecology, Economics, and Culture.* Princeton, NJ: Princeton University Press, 2005.

Boroditsky, L. "Linguistic Relativity." In *Encyclopedia of Cognitive Science,* edited by L. Nadel. London: Macmillan, 2003.

Boucher, D. H. "The Idea of Mutualism, Past and Future." In *The Biology of*

Mutualism: Ecology and Evolution, edited by D. H. Boucher. London: Croom Helm, 1986.

———. "Newtonian Ecology and Beyond." *Science as Culture* 7:493–517, 1998.

Bowers, C. A. *Education, Cultural Myths, and the Ecological Crisis: Toward Deep Changes.* Albany: State University of New York Press, 1993.

———. "How Language Limits Our Understanding of Environmental Education." *Environmental Education Research* 7:141–151, 2001.

Bowker, G. C., and S. L. Star. *Sorting Things Out: Classification and Its Consequences.* Cambridge, MA: MIT Press, 1999.

Bowler, P. J. "Social Metaphors in Evolutionary Biology, 1870–1930: The Wider Dimension of Social Darwinism." In *Biology as Society, Society as Biology: Metaphors,* edited by S. Maasen, E. Mendelsohn, and P. Weingart. Dordrecht: Kluwer, 1995.

———. *Life's Splendid Drama: Evolutionary Biology and the Reconstruction of Life's Ancestry, 1860–1940.* Chicago: University of Chicago Press, 1996.

Boyd, R. "Metaphor and Theory Change: What Is 'Metaphor' a Metaphor for?" In *Metaphor and Thought,* edited by A. Ortony, 2nd ed. Cambridge: Cambridge University Press, 1993.

Bradshaw, G. A., and M. Bekoff. "Ecology and Social Responsibility: The Re-Embodiment of Science." *Trends in Ecology & Evolution* 16:460–465, 2001.

Brand, F. S., and K. Jax. "Focusing the Meaning(s) of Resilience: Resilience as a Descriptive Concept and a Boundary Object." *Ecology and Society* 12, 2007, www.ecologyandsociety.org/vol12/iss1/art23.

Brown, T. L. *Making Truth: Metaphor in Science.* Urbana: University of Illinois Press, 2003.

Brunson, M. W. "Managing Naturalness as a Continuum: Setting Limits of Acceptable Change." In *Restoring Nature: Perspectives from the Social Sciences and Humanities,* edited by P. H. Gobster and R. B. Hull. Washington, DC: Island Press, 2000.

Buber, M. *I and Thou,* translated by R. G. Smith, 2nd ed. London: Continuum, 2004.

Bunnell, P. "Attributing Nature with Justifications." *Systems Research and Behavioral Science* 17:469–480, 2000.

Bury, J. B. *The Idea of Progress: An Inquiry into Its Origin and Growth.* New York: Macmillan, 1932.

Buttimer, A. *Geography and the Human Spirit.* Baltimore: Johns Hopkins University Press, 1993.

Caillon, S., and P. Degeorges. "Biodiversity: Negotiating the Border between Nature and Culture." *Biodiversity and Conservation* 16:2919–2931, 2007.

Callicott, J. B. "Multicultural Environmental Ethics." *Daedalus* 130:77–97, 2001.

Callicott, J. B., L. B. Crowder, and K. Mumford. "Current Normative Concepts in Conservation." *Conservation Biology* 13:22–35, 1999.

Callon, M. "Some Elements of a Sociology of Translation: Domestication of the Scallops and the Fishermen of St. Brieuc Bay." In *Power, Action, and Belief: A New Sociology of Knowledge*, edited by J. Law. New York: Routledge, 1986.

Campbell, D. T. "On the Conflicts between Biological and Social Evolution and between Psychology and Moral Tradition." *American Psychologist* 30:1103–1126, 1975.

Capra, F. *The Web of Life: A New Scientific Understanding of Living Systems.* New York: Doubleday, 1996.

Carew, A., and C. Mitchell. "Metaphors Used by Some Engineering Academics in Australia for Understanding and Explaining Sustainability." *Environmental Education Research* 12:217–231, 2006.

Carolan, M. S. "Society, Biology, and Ecology: Bringing Nature Back into Sociology's Disciplinary Narrative through Critical Realism." *Organization & Environment* 18:393–421, 2005.

———. "Science, Expertise, and the Democratization of the Decision-Making Process." *Society & Natural Resources* 19:661–668, 2006.

———. "The Values and Vulnerabilities of Metaphors within the Environmental Sciences." *Society & Natural Resources* 19:921–930, 2006.

———. "'This Is Not a Biodiversity Hotspot': The Power of Maps and Other Images in the Environmental Sciences." *Society & Natural Resources* 22:278–286, 2009.

Carpenter, S., B. Walker, J. M. Anderies, and N. Abel. "From Metaphor to Measurement: Resilience of What to What?" *Ecosystems* 4:765–781, 2001.

Cartwright, N. *The Dappled World: A Study of the Boundaries of Science.* Cambridge: Cambridge University Press, 1999.

Caudill, E. *Darwinian Myths: The Legends and Misuses of a Theory.* Knoxville: University of Tennessee Press, 1997.

Chaisson, E. J. "Ethical Evolution." *Zygon* 34:265–271, 1999.

———. *Cosmic Evolution: The Rise of Complexity in Nature.* Cambridge, MA: Harvard University Press, 2001.

Chawla, S. "Linguistic and Philosophical Roots of Our Environmental Crisis." *Environmental Ethics* 13:253–273, 1991.

Chew, M. K., and M. D. Laubichler. "Natural Enemies—Metaphor or Misconception?" *Science* 301:52–53, 2003.

Chilton, P. A. *Security Metaphors: Cold War Discourse from Containment to Common House.* New York: Peter Lang, 1996.

Cilliers, P. *Complexity and Postmodernism: Understanding Complex Systems.* New York: Routledge, 1998.

Clergeau, P., and M. A. Nuñez. "The Language of Fighting Invasive Species." *Science* 311:951, 2006.

Coates, P. *American Perceptions of Immigrant and Invasive Species: Strangers on the Land.* Berkeley and Los Angeles: University of California Press, 2008.

Cobern, W. W. "Science Education as an Exercise in Foreign Affairs." *Science & Education* 4:287–302, 1995.

Colautti, R. I., and H. J. MacIsaac. "A Neutral Terminology to Define 'Invasive' Species." *Diversity and Distributions* 10:135–141, 2004.

Colautti, R. I., and D. M. Richardson. "Subjectivity and Flexibility in Invasion Terminology: Too Much of a Good Thing?" *Biological Invasions* 11:1225–1229, 2009.

Collins, C. A., and S. R. Kephart. "Science as News: The Emergence and Framing of Biodiversity." *Mass Communication Review* 22:21–45, 1995.

Collins, H., and R. Evans. "The Third Wave of Science Studies: Studies of Expertise and Experience." *Social Studies of Science* 32:235–296, 2002.

Condit, C. M. "The Meaning and Effects of Discourse about Genetics: Methodological Variations in Studies of Discourse and Social Change." *Discourse & Society* 15:391–408, 2004.

Condit, C. M., T. Dubriwny, J. Lynch, and R. Parrott. "Lay People's Understanding of and Preference against the Word 'Mutation.'" *American Journal of Medical Genetics* 130A:245–250, 2004.

Cook, G. *Genetically Modified Language: The Discourse of Arguments for GM Crops and Food.* New York: Routledge, 2004.

Cooper, G. "Must There be a Balance of Nature?" *Biology and Philosophy* 16:481–506, 2001.

Coyne, R. *Designing Information Technology in the Postmodern Age: From Method to Metaphor.* Cambridge, MA: MIT Press, 1995.

Crist, E. "Against the Social Construction of Nature and Wilderness." *Environmental Ethics* 26:5–24, 2004.

Cronon, W. "The Trouble with Wilderness; or, Getting Back to the Wrong Nature." In *Uncommon Ground: Toward Reinventing Nature,* edited by W. Cronon. New York: W. W. Norton, 1995.

Crook, P. *Darwinism, War and History.* Cambridge: Cambridge University Press, 1994.

Csikszentmihalyi, M. *The Evolving Self: A Psychology for the Third Millennium.* New York: HarperCollins, 1993.

——. "The Mythic Potential of Evolution." *Zygon* 35:25–38, 2000.

Cuddington, K. "The 'Balance of Nature' Metaphor and Equilibrium in Population Ecology." *Biology and Philosophy* 16:463–479, 2001.

Dalai Lama. "Understanding Our Fundamental Nature." In *Visions of Compassion: Western Scientists and Tibetan Buddhists Examine Human Nature,* edited by R. J. Davidson and A. Harrington. New York: Oxford University Press, 2002.

Daston, L. "How Nature Became the Other: Anthropomorphism and Anthropocentrism in Early Modern Natural Philosophy." In *Biology as Society, Society as Biology: Metaphors,* edited by S. Maasen, E. Mendelsohn, and P. Weingart. Dordrecht: Kluwer, 1995.

Dauvergne, P. *The Shadows of Consumption: Consequences for the Global Environment.* Cambridge, MA: MIT Press, 2008.

Davis, M. A., and L. B. Slobodkin. "The Science and Values of Restoration Ecology." *Restoration Ecology* 12:1–3, 2004.

Davis, M. A., K. Thompson, and J. P. Grime. "Charles S. Elton and the Dissociation of Invasion Ecology from the Rest of Ecology." *Diversity and Distributions* 7:97–102, 2001.

Davis, W. *The Wayfinders: Why Ancient Wisdom Matters in the Modern World.* Toronto: Canadian Broadcasting Corporation, 2009.

Dawkins, R. *The Selfish Gene,* 2nd ed. New York: Oxford University Press, 1989.

de Carvalho, M. R., F. A. Bockmann, D. S. Amorim, and C. R. F. Brandao. "Systematics Must Embrace Comparative Biology and Evolution, Not Speed and Automation." *Evolutionary Biology* 35: 150–157, 2008.

de Waal, F. "Reading Nature's Tea Leaves." *Natural History,* December 2000/January 2001.

Dedaic, M. N., and D. N. Nelson, eds. *At War with Words.* New York: Mouton de Gruyter, 2003.

Demeritt, D. "Science Studies, Climate Change and the Prospects for Constructivist Critique." *Economy and Society* 35:453–479, 2006.

Dickinson, A. "Lyric Ethics: Ecocriticism, Material Metaphoricity, and the Poetics of Don McKay and Jan Zwicky." *Canadian Poetry* 55:34–52, 2004.

Dietz, T., and P. C. Stern. "Science, Values, and Biodiversity." *BioScience* 48:441–444, 1998.

Dirven, R., R. Frank, and C. Ilie, eds. *Language and Ideology,* vol. 2, *Descriptive Cognitive Approaches.* Amsterdam: John Benjamins, 2001.

Dirven, R., B. Hawkins, and E. Sandikcioglu, eds. *Language and Ideology,* vol. 1, *Theoretical Cognitive Approaches*. Amsterdam: John Benjamins, 2001.

Dorries, M., ed. *Experimenting in Tongues: Studies in Language and Science.* Stanford, CA: Stanford University Press, 2002.

Douglas, H. E. "The Moral Responsibilities of Scientists (Tensions between Autonomy and Responsibility)." *American Philosophical Quarterly* 40:59–68, 2003.

———. "Inserting the Public into Science." In *Democratization of Expertise? Exploring Novel Forms of Scientific Advice in Political Decision-Making,* edited by S. Maasen and P. Weingart. New York: Springer, 2005.

———. *Science, Policy, and the Value-Free Ideal.* Pittsburgh: University of Pittsburgh Press, 2009.

Duncan, S. L. "Mapping Whose Reality? Geographic Information Systems (GIS) and 'Wild Science.'" *Public Understanding of Science* 15:411–434, 2006.

Dupré, J. *The Disorder of Things: Metaphysical Foundations of the Disunity of Science.* Cambridge, MA: Harvard University Press, 1993.

———. "Fact and Value." In *Value-Free Science? Ideals and Illusions,* edited by H. Kincaid, J. Dupré, and A. Wylie. New York: Oxford University Press, 2007.

Earley, J. *Transforming Human Culture: Social Evolution and the Planetary Crisis.* Albany: State University of New York Press, 1997.

Ebach, M. C., and C. Holdrege. "More Taxonomy, Not DNA Barcoding." *BioScience* 55:822–823, 2005.

Eden, S. "Public Participation in Environmental Policy: Considering Scientific, Counter-Scientific and Non-Scientific Contributions." *Public Understanding of Science* 5:183–204, 1996.

Edge, D. "Technological Metaphor and Social Control." *New Literary History* 6:135–147, 1974.

Edwards, P. N. *The Closed World: Computers and the Politics of Discourse in Cold War America.* Cambridge, MA: MIT Press, 1996.

Ehrenfeld, J. R. "Industrial Ecology: Paradigm Shift or Normal Science?" *American Behavioral Scientist* 44:229–244, 2000.

———. *Sustainability by Design: A Subversive Strategy for Transforming Our Consumer Culture.* New Haven: Yale University Press, 2008.

Ehrlich, P. R. "Intervening in Evolution: Ethics and Actions." *Proceedings of the National Academy of Science* 98:5477–5480, 2001.

Ehrlich, P. R., and A. H. Ehrlich. *Extinction: The Causes and Consequences of the Disappearance of Species.* New York: Random House, 1981.

Einarsson, N. "All Animals Are Equal but Some Are Cetaceans: Conservation and Culture Conflict." In *Environmentalism: The View from Anthropology,* edited by K. Milton. New York: Routledge, 1993.

Einsiedel, E. "Public Engagement and Dialogue: A Research Review." In *Handbook of Public Communication of Science and Technology,* edited by M. Bucchi and B. Smart. New York: Routledge, 2008.

Elliott, K. C. "The Ethical Significance of Language in the Environmental Sciences: Case Studies from Pollution Research." *Ethics, Place and Environment* 12:157–173, 2009.

Ellis, R., C. Waterton, and B. Wynne. "Taxonomy, Biodiversity and Their Publics in Twenty-first-Century DNA Barcoding." *Public Understanding of Science* 19:497–512, 2010.

Entman, R. M. "Framing—Toward Clarification of a Fractured Paradigm." *Journal of Communication* 43:51–58, 1993.

Epstein, S. *Impure Science: AIDS, Activism, and the Politics of Knowledge.* Berkeley and Los Angeles: University of California Press, 1996.

Eubanks, P. *A War of Words in the Discourse of Trade: The Rhetorical Constitution of Metaphor.* Carbondale: Southern Illinois University Press, 2000.

Evans, J. M., A. C. Wilkie, and J. Burkhardt. "Adaptive Management of Nonnative Species: Moving beyond the 'Either-Or' through Experimental Pluralism." *Journal of Agricultural and Environmental Ethics* 21:521–539, 2008.

Evans, P. "Biodiversity: Nature for Nerds?" *Ecos* 17:7–12, 1996.

Evernden, N. *The Natural Alien: Humankind and Environment,* 2nd ed. Toronto: University of Toronto Press, 1993.

———. "Beyond Ecology: Self, Place, and the Pathetic Fallacy." In *The Ecocriticism Reader: Landmarks in Literary Ecology,* edited by C. Glotfelty and H. Fromm. Athens: University of Georgia Press, 1996.

Ewel, J. J., and F. E. Putz. "A Place for Alien Species in Ecosystem Restoration." *Frontiers in Ecology and the Environment* 7:354–360, 2004.

Fahnestock, J. *Rhetorical Figures in Science.* New York: Oxford University Press, 1999.

Fairclough, N., ed. *Critical Language Awareness.* New York: Longman, 1992.

Farber, P. L. *The Temptations of Evolutionary Ethics.* Berkeley and Los Angeles: University of California Press, 1994.

Feyerabend, P. "Quantum Theory and Our View of the World." In *Physics and Our View of the World,* edited by J. Hilgevoord. Cambridge: Cambridge University Press, 1994.

———. *Conquest of Abundance: A Tale of Abstraction versus the Richness of Being.* Chicago: University of Chicago Press, 1999.

Fill, A., and P. Mühlhäusler, eds. *The Ecolinguistics Reader: Language, Ecology and Environment.* New York: Continuum, 2001.

Fine, G. A., and L. Christoforides. "Dirty Birds, Filthy Immigrants, and the English Sparrow War: Metaphorical Linkage in Constructing Social Problems." *Symbolic Interaction* 14:375–393, 1991.

Fine, G. A., and K. Sandstrom. "Ideology in Action: A Pragmatic Approach to a Contested Concept." *Sociological Theory* 11:21–38, 1993.

Fiorino, D. J. "Citizen Participation and Environmental Risk: A Survey of Institutional Mechanisms." *Science, Technology and Human Values* 15:226–243, 1990.

Fischer, A., and R. van der Wal. "Invasive Plant Suppresses Charismatic Seabird: The Construction of Attitudes towards Biodiversity Management Options." *Biological Conservation* 135:256–267, 2007.

Fitzhugh, K. "DNA Barcoding: An Instance of Technology-Driven Science?" *BioScience* 56:462–463, 2006.

Fleck, L. *Genesis and Development of a Scientific Fact.* Chicago: University of Chicago Press, 1936.

Fleischner, T. L. "Natural History and the Deep Roots of Resource Management." *Natural Resources Journal* 45:1–13, 2005.

Fleming, P. A. "Can Nature (Legitimately) Be Our Guide?" In *Religion and the New Ecology: Environmental Responsibility in a World in Flux,* edited by D. M. Lodge and C. Hamlin. Notre Dame, IN: University of Notre Dame Press, 2006.

Foote, L., N. Krogman, and J. Spence. "Should Academics Advocate on Environmental Issues?" *Society & Natural Resources* 22:579–589, 2009.

Foster, J. "Making Sense of Stewardship: Metaphorical Thinking and the Environment." *Environmental Education Research* 11:25–36, 2005.

Foster, J., and L. A. Sandberg. "Friends or Foe? Invasive Species and Public Green Space in Toronto." *Geographical Review* 94:178–198, 2004.

Fowler, J. W. *Stages of Faith.* New York: Harper and Row, 1981.

Fracchia, J., and R. C. Lewontin. "Does Culture Evolve?" *History and Theory* 38:52–78, 1999.

Frank, R. M. "Shifting Identities: The Metaphorics of Nature-Culture Dualism in Western and Basque Models of Self," www.metaphorik .de/04/frank.htm, 2003.

Frank, R. M., R. Dirven, T. Ziemke, and E. Bernárdez, eds. *Body, Language and Mind,* vol. 2, *Sociocultural Situatedness.* New York: Mouton de Gruyter, 2008.

Frost, R. "Education by Poetry: A Meditative Monologue" (1931). In *The Collected Prose of Robert Frost,* edited by M. Richardson. Cambridge, MA: Harvard University Press, 2007.

Fuller, R. Buckminster. *Operating Manual for Spaceship Earth.* New York: Simon & Schuster, 1969.

Funtowicz, S. O., and J. R. Ravetz. "Science for the Post-Normal Age." *Futures* 25:739–755, 1993.

Gabriel, W., M. Lynch, and R. Burger. "Muller's Ratchet and Mutational Meltdowns." *Evolution* 47:1744–1757, 1993.

Galison, P., and D. J. Stump, eds. *The Disunity of Science: Boundaries, Contexts, and Power.* Stanford, CA: Stanford University Press, 1996.

Gaziano, E. "Ecological Metaphors as Scientific Boundary Work: Innovation and Authority in Interwar Sociology and Biology." *American Journal of Sociology* 101:874–907, 1996.

Geertz, C. *Local Knowledge.* New York: Basic Books, 1979.

Genet, R. M. "The Epic of Evolution: A Course Developmental Project." *Zygon* 33:635–644, 1998.

Gentner, D., and S. Goldin-Meadow, eds. *Language in Mind: Advances in the Study of Language and Thought.* Cambridge, MA: MIT Press, 2003.

Gibbons, M., C. Limoges, H. Nowotny, S. Schwartzman, P. Scott, and M. Trow. *The New Production of Knowledge: The Dynamics of Science and Research in Contemporary Societies.* Thousand Oaks, CA: Sage, 1994.

Gibbs, R. W., Jr. *The Poetics of Mind: Figurative Thought, Language, and Understanding.* Cambridge: Cambridge University Press, 1994.

———. "How Language Reflects the Embodied Nature of Creative Cognition." In *Creative Thought: An Investigation of Conceptual Structures and Processes,* edited by T. B. Ward, S. M. Smith, and J. Vaid. Washington, DC: American Psychological Association, 1997.

———. "Taking Metaphor Out of Our Heads and Putting It into the Cultural World." In *Metaphor in Cognitive Linguistics: Selected Papers from the Fifth International Cognitive Linguistics Conference,* edited by R. W. Gibbs Jr. and G. J. Steen. Amsterdam: John Benjamins, 1999.

Gibson, R. B. *Sustainability Assessment: Criteria and Processes.* London: Earthscan, 2005.

Gieryn, T. F. *Cultural Boundaries of Science: Credibility on the Line.* Chicago: University of Chicago Press, 1999.

Gilbert, S. F. "The Metaphorical Structuring of Social Perceptions." *Soundings* 62:166–186, 1979.

Glotfelty, C. "Cold War, Silent Spring: The Trope of War in Modern Environmentalism." In *And No Birds Sing: Rhetorical Analyses of Rachel Carson's Silent Spring,* edited by C. Waddell. Carbondale: Southern Illinois University Press, 2000.

Glover, R. "The War on _____." In *Collateral Language: A User's Guide to*

America's New War, edited by J. Collins and R. Glover. New York: New York University Press, 2002.

Goatly, A. "Green Grammar and Grammatical Metaphor, or Language and the Myth of Power, or Metaphors We Die By." *Journal of Pragmatics* 25:537–560, 1996.

———. *Washing the Brain: Metaphor and Hidden Ideology.* Amsterdam: John Benjamins, 2007.

Gobster, P. H. "Invasive Species as Ecological Threat: Is Restoration an Alternative to Fear-Based Resource Management?" *Ecological Restoration* 23:261–270, 2005.

Goldberg, N. *Writing Down the Bones: Freeing the Writer Within.* Boston: Shambhala, 1986.

Gould, S. J. *Ever Since Darwin: Reflections in Natural History.* New York: W. W. Norton, 1977.

———. *Wonderful Life: The Burgess Shale and the Nature of History.* New York: W. W. Norton, 1986.

———. "On Replacing the Idea of Progress with an Operational Notion of Directionality." In *Evolutionary Progress,* edited by M. H. Nitecki. Chicago: University of Chicago Press, 1988.

———. "Self-Help for a Hedgehog Stuck on a Molehill." *Evolution* 51:1020–1023, 1997.

Gray, B. "Framing of Environmental Disputes." In *Making Sense of Intractable Environmental Conflicts,* edited by R. J. Lewicki, B. Gray, and M. Elliott. Washington, DC: Island Press, 2003.

Greene, J. C. *Science, Ideology, and World View: Essays in the History of Evolutionary Ideas.* Berkeley and Los Angeles: University of California Press, 1981.

Gregory, J., and S. Miller. *Science in Public: Communication, Culture and Credibility.* New York: Plenum, 1998.

Gross, A. G. *The Rhetoric of Science.* Cambridge, MA: Harvard University Press, 1990.

Grove-White, R. "Environmentalism: A New Moral Discourse for Technological Society?" In *Environmentalism: The View from Anthropology,* edited by K. Milton. New York: Routledge, 1993.

Guha, R. "The Authoritarian Biologist and the Arrogance of Anti-Humanism: Wildlife Conservation in the Third World." *Ecologist* 27:14–20, 1997.

Gumperz, J. J., and S. C. Levinson, eds. *Rethinking Linguistic Relativity.* Cambridge: Cambridge University Press, 1996.

Gurevitch, J. "Commentary on Simberloff (2006): Meltdowns, Snowballs and Positive Feedbacks." *Ecology Letters* 9:919–921, 2006.

Gusfield, J. "The Literary Rhetoric of Science: Comedy and Pathos in Drinking Driver Research." *American Sociological Review* 41:16–34, 1976.

Gwyn, R. "'Captain of My Own Ship': Metaphor and the Discourse of Chronic Illness." In *Researching and Applying Metaphor*, edited by L. Cameron and G. Low. Cambridge: Cambridge University Press, 1999.

Hagen, J. B. *An Entangled Bank: Origins of Ecosystem Ecology.* New Brunswick, NJ: Rutgers University Press, 1992.

Haila, Y. "Biodiversity and the Divide between Culture and Nature." *Biodiversity and Conservation* 8:165–181, 1999.

———. "Beyond the Nature-Culture Dualism." *Biology and Philosophy* 15:155–175, 2000.

———. "A Conceptual Genealogy of Fragmentation Research: From Island Biogeography to Landscape Ecology." *Ecological Applications* 12:321–334, 2002.

Hall, S. J. "Cultural Disturbances and Local Ecological Knowledge Mediate Cattail (*Typha domingensis*) Invasion in Lake Pátzcuaro, México." *Human Ecology* 37:241–249, 2009.

Haraway, D. J. *Crystals, Fabrics, and Fields: Metaphors of Organicism in Twentieth-Century Developmental Biology.* New Haven: Yale University Press, 1976.

———. *Primate Visions: Gender, Race, and Nature in the World of Modern Science.* New York: Routledge, 1989.

———. *Modest_Witness@Second_Millennium. FemaleMan_Meets_Onco-Mouse: Feminism and Technoscience.* New York: Routledge, 1997.

Hardin, G. "The Tragedy of the Commons." *Science* 162:1243–1248, 1968.

Harding, S. "Should Philosophies of Science Encode Democratic Ideals?" In *Science, Technology, and Democracy,* edited by D. L. Kleinman. Albany: State University of New York Press, 2000.

Hardy-Short, D. C., and C. B. Short. "Fire, Death, and Rebirth: A Metaphoric Analysis of the 1988 Yellowstone Fire Debate." *Western Journal of Communication* 59:103–125, 1995.

Harman, W. W., and E. Sahtouris. *Biology Revisioned.* Berkeley: North Atlantic Books, 1998.

Harré, R., J. Brockmeier, and P. Mühlhäusler. *Greenspeak: A Study of Environmental Discourse.* Thousand Oaks, CA: Sage, 1999.

Harrington, A. "Metaphoric Connections: Holistic Science in the Shadow of the Third Reich." *Social Research* 62:357–385, 1995.

———. "A Science of Compassion or a Compassionate Science? What Do We Expect from a Cross-Cultural Dialogue with Buddhism?" In *Visions of Compassion: Western Scientists and Tibetan Buddhists Examine*

Human Nature, edited by R. J. Davidson and A. Harrington. New York: Oxford University Press, 2002.

Harrison, K. D. *When Languages Die: The Extinction of the World's Languages and the Erosion of Human Knowledge.* New York: Oxford University Press, 2007.

Hastings, G., M. Stead, and J. Webb. "Fear Appeals in Social Marketing: Strategic and Ethical Reasons for Concern." *Psychology & Marketing* 21:961–986, 2004.

Hawkins, M. J. "The Struggle for Existence in 19th-Century Social Theory: Three Case Studies." *History of the Human Sciences* 8:47–67, 1995.

———. *Social Darwinism in European and American Thought, 1860–1945: Nature as Model and Nature as Threat.* Cambridge: Cambridge University Press, 1997.

Hayles, N. K. "Desiring Agency: Limiting Metaphors and Enabling Constraints in Dawkins and Deleuze/Guattari." *Substance* 30:144–159, 2001.

Heberlein, T. A. "Wildlife Caretaking vs. Wildlife Management—A Short Lesson in Swedish." *Wildlife Society Bulletin* 33:1–3, 2005.

Hebert, P. D. N., and R. D. H. Barrett. "Reply to the Comment by L. Prendini on 'Identifying Spiders through DNA Barcodes.'" *Canadian Journal of Zoology* 83:505–506, 2005.

Hebert, P. D. N., A. Cywinska, S. L. Ball, and J. R. deWaard. "Biological Identifications through DNA Barcodes." *Proceedings of the Royal Society of London B—Biological Sciences* 270:313–321, 2003.

Hebert, P. D. N., and T. R. Gregory. "The Promise of DNA Barcoding for Taxonomy." *Systematic Biology* 54:852–859, 2005.

Hebert, P. D. N., M. Y. Stoeckle, T. S. Zemlak, and C. M. Francis. "Identification of Birds through DNA Barcodes." *PLoS Biology* 2, 2004, www.plosbiology.org/article/info:doi%2F10.1371%2Fjournal.pbio.0020312.

Heidegger, M. *The Question concerning Technology, and Other Essays.* New York: Garland, 1977.

Henderson, J. Y. "Ayukpachi: Empowering Aboriginal Thought." In *Reclaiming Indigenous Voice and Vision,* edited by M. Battiste. Vancouver: University of British Columbia Press, 2000.

Henrich, K. "Gaia Infiltrata: The Anthroposphere as a Complex Autoparasitic System." *Environmental Values* 11:489–507, 2002.

Herbers, J. M. "Watch Your Language! Racially Loaded Metaphors in Scientific Research." *BioScience* 57:104–105, 2007.

Hesse, M. "The Explanatory Function of Metaphor." In *Logic, Methodology and Philosophy of Science,* edited by Y. Bar-Hillel. Amsterdam: North-Holland, 1965.

———. *Revolutions and Reconstructions in the Philosophy of Science.* Bloomington: Indiana University Press, 1980.

———. "The Cognitive Claims of Metaphor." *Journal of Speculative Philosophy* 2:1–16, 1988.

Hilgartner, S. "The Dominant View of Popularization: Conceptual Problems, Political Uses." *Social Studies of Science* 20:519–539, 1990.

Hill, J. A. "'Expert Rhetorics' in Advocacy for Endangered Languages: Who Is Listening, and What Do They Hear?" *Journal of Linguistic Anthropology* 12:119–133, 2002.

Ho, M. W., and S. W. Fox, eds. *Evolutionary Processes and Metaphors.* New York: Wiley, 1988.

Hodge, M. J. S. "Natural Selection: Historical Perspectives." In *Keywords in Evolutionary Biology,* edited by E. F. Keller and E. A. Lloyd. Cambridge, MA: Harvard University Press, 1992.

Hodges, K. E. "Defining the Problem: Terminology and Progress in Ecology." *Frontiers in Ecology and the Environment* 6: 35–42, 2008.

Holland, E. M., A. Pleasant, S. Quatrano, R. Gerst, M. C. Nisbet, and C. Mooney. "Letters: The Risks and Advantages of Framing Science." *Science* 317:1168–1170, 2007.

Holloway, M. "Democratizing Taxonomy." *Conservation in Practice* 7:14–21, 2006.

Holt, A. "Biodiversity Definitions Vary within the Discipline." *Nature* 444:146, 2006.

Horkheimer, M., and T. W. Adorno. *Dialectic of Enlightenment,* translated by J. Cumming. New York: Herder and Herder, 1972.

Hubbard, B. M. *Conscious Evolution: Awakening the Power of Our Social Potential.* Novato, CA: New World Library, 1998.

Hull, R. B., D. Richert, E. Seekamp, D. Robertson, and G. J. Buhyoff. "Understandings of Environmental Quality: Ambiguities and Values Held by Environmental Professionals." *Environmental Management* 31:1–13, 2003.

Hull, R. B., and D. P. Robertson. "The Language of Nature Matters: We Need a More Public Ecology." In *Restoring Nature: Perspectives from the Social Sciences and Humanities,* edited by P. H. Gobster and R. B. Hull. Washington, DC: Island Press, 2000.

Hull, R. B., D. P. Robertson, D. Richert, E. Seekamp, and G. J. Buhyoff. "Assumptions about Ecological Scale and Nature Knowing Best Hiding in Environmental Decisions." *Conservation Ecology* 6, 2002, www .ecologyandsociety.org/vol6/iss2/art12.

Hulme, M. "The Conquering of Climate: Discourses of Fear and Their Dissolution." *Geographical Journal* 174:5–16, 2008.

Hurlbert, S. H. "Functional Importance vs Keystoneness: Reformulating Some Questions in Theoretical Biocenology." *Australian Journal of Ecology* 22:369–382, 1997.

Ingold, T. *The Perception of the Environment: Essays in Livelihood, Dwelling and Skill.* New York: Routledge, 2000.

Irwin, A., and B. Wynne, eds. *Misunderstanding Science? The Public Reconstruction of Science and Technology.* Cambridge: Cambridge University Press, 1996.

Isaac, N. J. B., J. Mallett, and G. M. Mace. "Taxonomic Inflation: Its Influence on Macroecology and Conservation." *Trends in Ecology & Evolution* 19:464–469, 2004.

Jackson, T. "Sustainability and the 'Struggle for Existence': The Critical Role of Metaphor in Society's Metabolism." *Environmental Values* 12:289–316, 2003.

Janks, H., and R. Ivanic. "Critical Language Awareness and Emancipatory Discourse." In *Critical Language Awareness,* edited by N. Fairclough. New York: Longman, 1992.

Janusz, S. "Feminism and Metaphor—Friend, Foe, Force." *Metaphor and Symbolic Activity* 9:289–300, 1994.

Janzen, D. H. "Now Is the Time." *Philosophical Transactions of the Royal Society of London B—Biological Sciences* 359:731–732, 2004.

Jasanoff, S. "Technologies of Humility: Citizen Participation in Governing Science." *Minerva* 41:223–244, 2003.

Jelinski, D. E. "There Is No Mother Nature—There Is No Balance of Nature: Culture, Ecology and Conservation." *Human Ecology* 33:271–288, 2005.

Jenkins, W. "Assessing Metaphors of Agency: Intervention, Perfection, and Care as Models of Environmental Practice." *Environmental Ethics* 27:135–154, 2005.

Journet, D. "Ecological Theories as Cultural Narratives: F. E. Clement's and H. A. Gleason's 'Stories' of Community Succession." *Written Communication* 8:446–472, 1991.

Kates, R., T. Parris, and A. Leiserowitz. "What Is Sustainable Development?" *Environment* 47:8–21, 2005.

Kauffman, S. A. *The Origins of Order: Self-Organization and Selection in Evolution.* New York: Oxford University Press, 1993.

Kay, L. E. *Who Wrote the Book of Life? A History of the Genetic Code.* Stanford, CA: Stanford University Press, 2000.

Keddy, P. A. *Competition.* New York: Chapman and Hall, 1989.

Keller, E. F. "Demarcating Public from Private Values in Evolutionary Discourse." *Journal of the History of Biology* 21:195–211, 1988.

———. "Language and Ideology in Evolutionary Theory: Reading Cultural

Norms into Natural Law." In *The Boundaries of Humanity: Humans, Animals, Machines,* edited by J. J. Sheehan and M. Sosna. Berkeley and Los Angeles: University of California Press, 1991.

———. "Competition: Current Usages." In *Keywords in Evolutionary Biology,* edited by E. F. Keller and E. A. Lloyd. Cambridge, MA: Harvard University Press, 1992.

———. *Refiguring Life: Metaphors of Twentieth-Century Biology.* New York: Columbia University Press, 1995.

———. *Making Sense of Life: Explaining Biological Development with Models, Metaphors, and Machines.* Cambridge, MA: Harvard University Press, 2002.

Keller, E. F., and E. A. Lloyd. "Introduction." In *Keywords in Evolutionary Biology,* edited by E. F. Keller and E. A. Lloyd. Cambridge, MA: Harvard University Press, 1992.

Keulartz, J., and C. van der Weele. "Framing and Reframing in Invasion Biology." *Configurations* 16:93–115, 2008.

Kevles, D. J. "Eugenics." In *Keywords in Evolutionary Biology,* edited by E. F. Keller and E. A. Lloyd. Cambridge, MA: Harvard University Press, 1992.

Killingsworth, M. J., and J. S. Palmer. *Ecospeak: Rhetoric and Environmental Politics in America.* Carbondale: Southern Illinois University Press, 1992.

Kincaid, H., J. Dupré, and A. Wylie, eds. *Value-Free Science? Ideals and Illusions.* New York: Oxford University Press, 2007.

Kinchy, A., and D. L. Kleinman. "Organizing Credibility: Discursive and Organizational Orthodoxy on the Borders of Ecology and Politics." *Social Studies of Science* 33:869–896, 2003.

Kirmayer, L. J. "The Body's Insistence on Meaning: Metaphor as Presentation and Representation in Illness Experience." *Medical Anthropology Quarterly* 6:323–346, 1992.

Kitcher, P. *Science, Truth, and Democracy.* New York: Oxford University Press, 2001.

———. "Responsible Biology." *BioScience* 54:331–336, 2004.

Klamer, A., and T. C. Leonard. "So What's an Economic Metaphor?" In *Natural Images in Economic Thought: "Markets Read in Tooth and Claw,"* edited by P. Mirowski. Cambridge: Cambridge University Press, 1994.

Kleinman, D. L. "Beyond the Science Wars: Contemplating the Democratization of Science." *Politics and the Life Sciences* 16:133–145, 1998.

Knoll, A. H., and R. K. Bambach. "Directionality in the History of Life: Diffusion from the Left Wall or Repeated Scaling of the Right?" *Paleobiology* 26:1–14, 2000.

Knudsen, S. "Scientific Metaphors Going Public." *Journal of Pragmatics* 35:1247–1263, 2003.

Kolodny, A. *The Lay of the Land: Metaphor as Experience and History in American Life and Letters.* Chapel Hill: University of North Carolina Press, 1975.

Kuhn, T. S. *The Structure of Scientific Revolutions,* 2nd ed. Chicago: University of Chicago Press, 1970.

———. "Metaphor in Science." In *Metaphor and Thought,* edited by A. Ortony, 2nd ed. Cambridge: Cambridge University Press, 1993.

Kwa, C. "Representation of Nature Mediating between Ecology and Science Policy: The Case of the International Biological Program." *Social Studies of Science* 17:413–442, 1987.

Lach, D., P. List, B. Steel, and B. Shindler. "Advocacy and Credibility of Ecological Scientists in Resource Decisionmaking: A Regional Study." *BioScience* 53:170–178, 2003.

Lackey, R. T. "Values, Policy, and Ecosystem Health." *BioScience* 51:437–443, 2001.

———. "Science, Scientists, and Policy Advocacy." *Conservation Biology* 21:12–17, 2007.

Lakoff, G. *Don't Think of an Elephant! Know Your Values and Frame the Debate—The Essential Guide for Progressives.* White River Junction, VT: Chelsea Green, 2004.

Lakoff, G., and M. Johnson. *Metaphors We Live By.* Chicago: University of Chicago Press, 1980.

———. *Philosophy in the Flesh: The Embodied Mind and Its Challenge to Western Thought.* New York: Basic Books, 1999.

Larson, B. M. H. "The War of the Roses: Demilitarizing Invasion Biology." *Frontiers in Ecology and the Environment* 3:495–500, 2005.

———. "The Social Resonance of Competitive and Progressive Evolutionary Metaphors." *BioScience* 56:997–1004, 2006.

———. "An Alien Approach to Invasive Species: Objectivity and Society in Invasion Biology." *Biological Invasions* 9:947–956, 2007.

———. "DNA Barcoding: The Social Frontier." *Frontiers in Ecology and the Environment* 5:437–442, 2007.

———. "Thirteen Ways of Looking at Invasive Species." In *Invasive Plants: Inventories, Strategies, and Action,* edited by D. R. Clements and S. Darbyshire. Sainte Anne de Bellevue, QC: Canadian Weed Science Society/Société Canadienne de Malherbologie, 2007.

———. "Who's Invading What? Systems Thinking about Invasive Species." *Canadian Journal of Plant Sciences* 87:993–999, 2007.

———. "Entangled Biological, Cultural, and Linguistic Origins of the War on Invasive Species." In *Body, Language and Mind*, vol. 2, *Sociocultural Situatedness*, edited by R. Frank, R. Dirven, T. Ziemke, and E. Bernárdez. New York: Mouton de Gruyter, 2008.

—. "Embodied Realism and Invasive Species." In *Philosophy of Ecology and Conservation Biology*, edited by K. de Laplante and K. Peacock. New York: Elsevier, forthcoming.

Latour, B. *We Have Never Been Modern*. Cambridge, MA: Harvard University Press, 1993.

—. *Politics of Nature: How to Bring the Sciences into Democracy.* Cambridge, MA: Harvard University Press, 2004.

Latour, B., and S. Woolgar. "Laboratory Life: The Construction of Scientific Facts." Beverly Hills, CA: Sage, 1979.

Lenoir, T., ed. *Inscribing Science: Scientific Texts and the Materiality of Communication*. Stanford, CA: Stanford University Press, 1998.

Lewicki, R. J., B. Gray, and M. Elliott, eds. *Making Sense of Intractable Environmental Conflicts*. Washington, DC: Island Press, 2003.

Lewontin, R. C. *Biology as Ideology: The Doctrine of DNA*. Concord, ON: House of Anansi Press, 1991.

Livingston, I. *Between Science and Literature: An Introduction to Autopoetics*. Urbana: University of Illinois Press, 2006.

Livingstone, D. N., and R. T. Harrison. "Meaning through Metaphor: Analogy as Epistemology." *Annals of the Association of American Geographers* 71:95–107, 1981.

Locke, S. "Golem Science and the Public Understanding of Science: From Deficit to Dilemma." *Public Understanding of Science* 8:75–92, 1999.

Lodge, D. M., and C. Hamlin, eds. *Religion and the New Ecology: Environmental Responsibility in a World in Flux*. Notre Dame, IN: University of Notre Dame Press, 2006.

Lonergan, B. J. F. *Insight: A Study of Human Understanding*. London: Longmans, Green, 1958.

Longino, H. E. *Science as Social Knowledge: Values and Objectivity in Scientific Inquiry*. Princeton, NJ: Princeton University Press, 1990.

—. "Gender and Racial Biases in Scientific Research." In *Ethics of Scientific Research*, edited by K. Shrader-Frechette. Lanham, MD: Rowman and Littlefield, 1994.

López, J. J. "Notes on Metaphors, Notes as Metaphors: The Genome as Musical Spectacle." *Science Communication* 29:7–34, 2007.

Louv, R. *Last Child in the Woods: Saving Our Children from Nature-Deficit Disorder*. Chapel Hill, NC: Algonquin Books, 2008.

Lovelock, J. *Gaia: A New Look at Life on Earth*. New York: Oxford University Press, 1979.

Low, S. M. "Embodied Metaphors: Nerves as Lived Experience." In
    Embodiment and Experience: The Existential Ground of Culture and Self,
    edited by T. J. Csordas. Cambridge: Cambridge University Press, 1994.
Lubchenco, J. "Entering the Century of the Environment: A New Social
    Contract for Science." *Science* 279:491–497, 1998.
Maasen, S., E. Mendelsohn, and P. Weingart, eds. *Biology as Society, Society
    as Biology: Metaphors.* Dordrecht: Kluwer, 1995.
Maasen, S., and P. Weingart. "Metaphors—Messengers of Meaning:
    A Contribution to an Evolutionary Sociology of Science." *Science
    Communication* 17:9–31, 1995.
MacDougall, A. S., and R. Turkington. "Are Invasive Species the Drivers
    or Passengers of Change in Degraded Ecosystems?" *Ecology* 86:42–55,
    2005.
Mackenzie, B. F., and B. M. H. Larson. "Participation under Time Con-
    straints: Landowner Perceptions of Rapid Response to the Emerald Ash
    Borer." *Society & Natural Resources* 23: 1013–1022, 2010.
Macnaughten, P., and J. Urry. *Contested Natures.* Thousand Oaks, CA: Sage,
    1998.
Macy, J. *Mutual Causality in Buddhism and General Systems Theory: The
    Dharma of Living Systems.* Albany: State University of New York Press,
    1991.
Maguire, J. "The Tears Inside the Stone: Reflections on the Ecology of Fear."
    In *Risk, Environment and Modernity: Towards a New Ecology,* edited by
    S. Lash, B. Szerszynski, and B. Wynne. Thousand Oaks, CA: Sage, 1996.
Maguire, L. A. "What Can Decision Analysis Do for Invasive Species
    Management?" *Risk Analysis* 24:859–868, 2004.
Maienschein, J., and M. Ruse, eds. *Biology and the Foundation of Ethics.*
    Cambridge: Cambridge University Press, 1999.
Margulis, L. "Words as Battle Cries—Symbiogenesis and the New Field
    of Endocytobiology." *BioScience* 40:673–677, 1990.
Marshall, P. *Nature's Web: Rethinking Our Place on Earth.* New York:
    Paragon House, 1992.
Martin, E. "Toward an Anthropology of Immunology: The Body as Nation
    State." *Medical Anthropology Quarterly* 4:410–426, 1990.
———. "The Egg and the Sperm: How Science Has Constructed a Romance
    Based on Stereotypical Male-Female Roles." *Signs* 16:485–501, 1991.
Marx, L., and B. Mazlish, eds. *Progress: Fact or Illusion?* Ann Arbor:
    University of Michigan Press, 1996.
Maslow, A. H. *Toward a Psychology of Being,* 2nd ed. New York: Van
    Nostrand, 1968.
Maynard Smith, J. "Evolutionary Progress and Levels of Selection." In

Evolutionary Progress, edited by M. H. Nitecki. Chicago: University of Chicago Press, 1988.

McInerney, J. O., J. A. Cotton, and D. Pisani. "The Prokaryotic Tree of Life: Past, Present . . . and Future?" *Trends in Ecology & Evolution* 23:276–281, 2008.

McIntosh, R. "Competition: Historical Perspectives." In *Keywords in Evolutionary Biology,* edited by E. F. Keller and E. A. Lloyd. Cambridge, MA: Harvard University Press, 1992.

McKenzie-Mohr, D., and W. Smith. *Fostering Sustainable Behavior: An Introduction to Community-Based Social Marketing.* Gabriola Island, BC: New Society, 1999.

McMichael, T. "Fine Battlefield Reporting, but It's Time to Stop the War Metaphor." *Science* 295:1469, 2002.

McShea, D. W. "Metazoan Complexity and Evolution: Is There a Trend?" *Evolution* 50:477–492, 1996.

Medawar, P. B. "Review of the *Phenomenon of Man.*" *Mind* 70:99–106, 1961.

Medley, K. E., and H. W. Kalibo. "Global Localism: Recentering the Research Agenda for Biodiversity Conservation." *Natural Resources Forum* 31:151–161, 2007.

Meiners, S. J. "Native and Exotic Plant Species Exhibit Similar Population Dynamics during Succession." *Ecology* 88:1098–1104, 2007.

Meisner, M. "Metaphors of Nature: Old Vinegar in New Bottles?" *Trumpeter* 12:11–18, 1995, http://trumpeter.athabascau.ca/index.php/trumpet/article/view/845/1218.

Menaker, E., and W. Menaker. *Ego in Evolution.* New York: Grove Press, 1965.

Merchant, C. *The Death of Nature: Women, Ecology, and the Scientific Revolution.* New York: Harper and Row, 1980.

———. "Partnership Ethics and Cultural Discourse: Women and the Earth Summit." In *Living with Nature: Environmental Politics as Cultural Discourse,* edited by F. Fischer and M. A. Hajer. New York: Oxford University Press, 1999.

———. *Radical Ecology: The Search for a Livable World,* 2nd ed. New York: Routledge, 2005.

Meyer, C. P., and G. Paulay. "DNA Barcoding: Error Rates Based on Comprehensive Sampling." *PLoS Biology* 3, 2005, www.plosbiology.org/article/info:doi/10.1371/journal.pbio.0030422.

Midgley, M. *Science as Salvation: A Modern Myth and Its Meaning.* New York: Routledge, 1992.

———. *The Myths We Live By.* New York: Routledge, 2003.

————, ed. *Earthly Realism: The Meaning of Gaia.* Charlottesville, VA:
    Imprint Academic, 2007.

Mikkelson, G. M. "Methods and Metaphors in Community Ecology: The
    Problem of Defining Stability." *Perspectives on Science* 5:481–498, 1997.

Miller, J. R. "Biodiversity Conservation and the Extinction of Experience."
    Trends in Ecology & Evolution 20:430–434, 2005.

Mills, W. J. "Metaphorical Vision: Changes in Western Attitudes to the
    Environment." *Annals of the Association of American Geographers*
    72:237–253, 1982.

Milton, K. "Ducks Out of Water: Nature Conservation as Boundary Main-
    tenance." In *Natural Enemies: People–Wildlife Conflicts in Anthropologi-
    cal Perspective,* edited by J. Knight. New York: Routledge, 2000.

Minteer, B., J. Collins, and S. Bird. "Editors' Overview: The Emergence of
    Ecological Ethics." *Science and Engineering Ethics* 14:473–481, 2008.

Mio, J. S. "Metaphor and Politics." *Metaphor and Symbol* 12:113–133, 1997.

Mio, J. S., S. C. Thompson, and G. H. Givens. "The Commons Dilemma
    as Metaphor: Memory, Influence, and Implications for Environmental
    Conservation." *Metaphor and Symbolic Activity* 8:23–42, 1993.

Mitman, G. "From the Population to Society: The Cooperative Metaphors
    of W. C. Allee and A. E. Emerson." *Journal of the History of Biology*
    21:173–194, 1988.

Monod, J. *Chance and Necessity: An Essay on the Natural Philosophy of
    Modern Biology,* translated by A. Wainhouse. New York: Alfred A.
    Knopf, 1971.

Montgomery, S. L. *The Scientific Voice.* New York: Guilford Press, 1996.

————. "Of Towers, Walls, and Fields: Perspectives on Language in
    Science." *Science* 303:1333–1335, 2004.

Moore, J. "Socializing Darwinism: Historiography and the Fortunes of
    a Phrase." In *Science as Politics,* edited by L. Levidow. London: Free
    Association Books, 1986.

————. "Revolution of the Space Invaders: Darwin and Wallace on the
    Geography of Life." In *Geography and Revolution,* edited by D. N.
    Livingstone and C. W. J. Withers. Chicago: University of Chicago Press,
    2005.

Moser, S. C., and L. Dilling, eds. *Creating a Climate for Change: Communi-
    cating Climate Change and Facilitating Social Change.* Cambridge:
    Cambridge University Press, 2007.

Muirhead, J., B. Leung, C. van Overdijk, D. Kelly, K. Nandakumar,
    K. Marchant, and H. MacIsaac. "Modelling Local and Long-Distance
    Dispersal of Invasive Emerald Ash Borer *Agrilus planipennis* (Coleop-
    tera) in North America." *Diversity and Distributions* 12:71–79, 2006.

Murphy, P. "Sex-Typing the Planet: Gaia Imagery and the Problem of Subverting Patriarchy." *Environmental Ethics* 10:155–168, 1988.

Myers, G. *Writing Biology: Texts in the Social Construction of Scientific Knowledge.* Madison: University of Wisconsin Press, 1990.

Nabhan, G. P. *Cross-Pollinations: The Marriage of Science and Poetry.* Minneapolis: Milkweed, 2004.

Nabhan, G. P., and S. St. Antoine. "The Loss of Floral and Faunal Story: The Extinction of Experience." In *The Biophilia Hypothesis,* edited by S. R. Kellert and E. O. Wilson. Washington, DC: Island Press, 1995.

Nelkin, D. "Promotional Metaphors and Their Popular Appeal." *Public Understanding of Science* 3:25–31, 1994.

———. "Molecular Metaphors: The Gene in Popular Discourse." *Nature Reviews Genetics* 2:555–559, 2001.

Nelson, D. N. "Conclusion: Word Peace." In *At War with Words,* edited by M. N. Dedaic and D. N. Nelson. New York: Mouton de Gruyter, 2003.

Nelson, M. P., and J. A. Vucetich. "On Advocacy by Environmental Scientists: What, Whether, Why, and How." *Conservation Biology* 23:1090–1101, 2009.

Nerlich, B. "Tracking the Fate of the Metaphor Silent Spring in British Environmental Discourse: Towards an Evolutionary Ecology of Metaphor," www.metaphorik.de/04/nerlich.htm, 2003.

Neumann, R. P. "Moral and Discursive Geographies in the War for Biodiversity in Africa." *Political Geography* 23:813–837, 2004.

Nisbet, M. C. "The Ethics of Framing Science." In *Communicating Biological Sciences: Ethical and Metaphorical Dimensions,* edited by B. Nerlich, R. Elliott, and B. Larson. Burlington, VT: Ashgate, 2009.

Nisbet, M. C., and C. Mooney. "Framing Science." *Science* 316:56, 2007.

Nisbet, R. *History of the Idea of Progress.* New Brunswick, NJ: Transaction Publishers, 1994.

Nordgren, A. "Metaphors in Behavioral Genetics." *Theoretical Medicine* 24:59–77, 2003.

Nordhaus, T., and M. Shellenberger. *Break Through: From the Death of Environmentalism to the Politics of Possibility.* New York: Houghton Mifflin, 2007.

Norgaard, K. M. "The Politics of Invasive Weed Management: Gender, Race, and Risk Perception in Rural California." *Rural Sociology* 72:450–477, 2007.

Norton, B. G. "Improving Ecological Communication: The Role of Ecologists in Environmental Policy Formation." *Ecological Applications* 8:350–364, 1998.

————. *Sustainability: A Philosophy of Adaptive Ecosystem Management.*
Chicago: University of Chicago Press, 2005.

————. "Beyond Positivist Ecology: Toward an Integrated Ecological
Ethics." *Science and Engineering Ethics* 14:581–592, 2008.

Norton, B. G., and D. Noonan. "Ecology and Valuation: Big Changes
Needed." *Ecological Economics* 63:664–675, 2007.

Novacek, M. J. "Engaging the Public in Biodiversity Issues." *Proceedings
of the National Academy of Science* 105:11571–11578, 2008.

O'Brien, W. "Exotic Invasions, Nativism, and Ecological Restoration:
On the Persistence of a Contentious Debate." *Ethics, Place and
Environment* 9:63–77, 2006.

Oelschlaeger, M. *Caring for Creation: An Ecumenical Approach to the
Environmental Crisis.* New Haven: Yale University Press, 1994.

Oliver, M. *The Leaf and the Cloud.* Cambridge, MA: Da Capo Press, 2000.

O'Neill, S., and S. Nicholson-Cole. "'Fear Won't Do It': Promoting Positive
Engagement with Climate Change through Visual and Iconic Representations." *Science Communication* 30:355–379, 2009.

Ono, K. A., and J. M. Sloop. *Shifting Borders: Rhetoric, Immigration, and
California's Proposition 187.* Philadelphia: Temple University Press, 2002.

Ornstein, R., and P. Ehrlich. *New World, New Mind: Moving toward Conscious Evolution.* New York: Doubleday, 1989.

Ortony, A. "Metaphor: A Multidimensional Problem." In *Metaphor and
Thought,* edited by A. Ortony, 2nd ed. Cambridge: Cambridge University Press, 1993.

Osowski, J. V. "Ensembles of Metaphor in the Psychology of William
James." In *Creative People at Work: Twelve Cognitive Case Studies,* edited
by D. B. Wallace and H. E. Gruber. New York: Oxford University Press,
1989.

Otis, L. *Membranes: Metaphors of Invasion in Nineteenth-Century Literature, Science, and Politics.* Baltimore: Johns Hopkins University Press,
1999.

————. "The Metaphoric Circuit: Organic and Technological Communication in the Nineteenth Century." *Journal of the History of Ideas* 63:105–
128, 2002.

Oyama, S. *Evolution's Eye: A Systems View of the Biology-Culture Divide.*
Durham, NC: Duke University Press, 2000.

Palmer, M. A., E. S. Bernhardt, E. A. Chornesky, S. L. Collins, A. P. Dob-
son, C. S. Duke, B. D. Gold, et al. "Ecological Science and Sustainability
for the 21st Century." *Frontiers in Ecology and the Environment* 3:4–11,
2005.

Paul, D. B. "Fitness: Historical Perspectives." In *Keywords in Evolutionary*

Biology, edited by E. F. Keller and E. A. Lloyd. Cambridge, MA: Harvard University Press, 1992.

———. "Darwin, Social Darwinism and Eugenics." In *The Cambridge Companion to Darwin,* edited by J. Hodge and G. Radick. Cambridge: Cambridge University Press, 2003.

Pauly, P. J. "The Beauty and Menace of the Japanese Cherry Trees: Conflicting Visions of American Ecological Independence." *Isis* 87:51–73, 1996.

Pawson, S. M., R. K. Didham, E. G. Brockerhoff, and E. D. Meenken. "Non-Native Plantation Forests as Alternative Habitat for Native Forest Beetles in a Heavily Modified Landscape." *Biodiversity and Conservation* 17:1127–1148, 2008.

Peat, D. *Blackfoot Physics: A Journey into the Native American Universe.* London: Fourth Estate, 1996.

Penman, R. "Environmental Matters and Communication Challenges." In *The Ecolinguistics Reader: Language, Ecology and Environment,* edited by A. Fill and P. Mühlhäusler. New York: Continuum, 2001.

Pepper, S. *World Hypotheses: A Study in Evidence.* Berkeley and Los Angeles: University of California Press, 1942.

Pergams, O. R. W., and P. A. Zaradic. "Is Love of Nature in the US Becoming Love of Electronic Media? 16-Year Downtrend in National Park Visits Explained by Watching Movies, Playing Video Games, Internet Use, and Oil Prices." *Journal of Environmental Management* 80:387–393, 2006.

Pfeiffer, J. M., and R. A. Voeks. "Biological Invasions and Biocultural Diversity: Linking Ecological and Cultural Systems." *Environmental Conservation* 35:281–293, 2008.

Philippon, D. J. *Conserving Words: How American Nature Writers Shaped the Environmental Movement.* Athens: University of Georgia Press, 2004.

Pickett, S. T. A., and M. L. Cadenasso. "The Ecosystem as a Multidimensional Concept: Meaning, Model, and Metaphor." *Ecosystems* 5:1–10, 2002.

Pickett, S. T. A., M. L. Cadenasso, and J. M. Grove. "Resilient Cities: Meaning, Models, and Metaphor for Integrating the Ecological, Socio-Economic, and Planning Realms." *Landscape and Urban Planning* 69:369–384, 2004.

Pielke, R. A. Jr. *The Honest Broker: Making Sense of Science in Policy and Politics.* Cambridge: Cambridge University Press, 2007.

Pigliucci, M. "Design Yes, Intelligent No: A Critique of Intelligent Design Theory and Neocreationism." *Skeptical Inquirer,* September–October 2001.

———. *Denying Evolution: Creationism, Scientism, and the Nature of Science.* Sunderland, MA: Sinauer, 2002.

Porteous, J. D. "Bodyscape: The Body-Landscape Metaphor." *Canadian Geographer* 30:2–12, 1986.

Proctor, J. D. "Expanding the Scope of Science and Ethics." *Annals of the Association of American Geographers* 88:290–296, 1998.

———. "Environment after Nature: Time for a New Vision." In *Envisioning Nature, Science, and Religion,* edited by J. D. Proctor. West Conshohocken, PA: Templeton Foundation Press, 2009.

Proctor, J. D., and B. M. H. Larson. "Ecology, Complexity, and Metaphor." *BioScience* 55:1065–1068, 2005.

Proctor, R. N. *Value-Free Science? Purity and Power in Modern Knowledge.* Cambridge, MA: Harvard University Press, 1991.

Proulx, S. R., D. E. L. Promislow, and P. C. Phillips. "Network Thinking in Ecology and Evolution." *Trends in Ecology & Evolution* 20:345–353, 2005.

Putnam, H. *The Collapse of the Fact/Value Dichotomy and Other Essays.* Cambridge, MA: Harvard University Press, 2002.

Quinn, N. "The Cultural Basis of Metaphor." In *Beyond Metaphor: The Theory of Tropes in Anthropology,* edited by J. W. Fernandez. Stanford, CA: Stanford University Press, 1991.

Rasmussen, B. "Poetic Truths and Clinical Reality: Client Experience of the Use of Metaphor by Therapists." *Smith College Studies in Social Work* 70:355–373, 2000.

Raup, D. M., and J. J. Sepkoski Jr. "Mass Extinctions in the Marine Fossil Record." *Science* 215:1501–1503, 1982.

Raven, P. H. "Taxonomy: Where Are We Now?" *Philosophical Transactions of the Royal Society of London B—Biological Sciences* 359:729–730, 2004.

Rawles, K. "Biological Diversity and Conservation Policy." In *Philosophy and Biodiversity,* edited by M. Oksanen and J. Pietarinen. Cambridge: Cambridge University Press, 2004.

Reddy, M. J. "The Conduit Metaphor: A Case of Frame Conflict in Our Language about Language." In *Metaphor and Thought,* edited by A. Ortony, 2nd ed. Cambridge: Cambridge University Press, 1993.

Rhoten, D., and A. Parker. "Risks and Rewards of an Interdisciplinary Research Path." *Science* 306:2046, 2004.

Ricciardi, A., and J. Cohen. "The Invasiveness of an Introduced Species Does Not Predict Its Impact." *Biological Invasions* 9:309–315, 2007.

Richards, I. A. *The Philosophy of Rhetoric.* London: Oxford University Press, 1936.

Richards, J. R. *Human Nature after Darwin: A Philosophical Introduction.* New York: Routledge, 2000.

Richards, R. J. "Evolution." In *Keywords in Evolutionary Biology,* edited by
    E. F. Keller and E. A. Lloyd. Cambridge, MA: Harvard University Press,
    1992.

————. *The Meaning of Evolution: The Morphological Construction and
    Ideological Reconstruction of Darwin's Theory.* Chicago: University of
    Chicago Press, 1992.

Ricoeur, P. *The Rule of Metaphor: Multi-Disciplinary Studies of the Creation
    of Meaning in Language,* translated by R. Czerny. Toronto: University of
    Toronto Press, 1977.

Ridder, B. "An Exploration of the Value of Naturalness and Wild Nature."
    Journal of Agricultural and Environmental Ethics 20:195–213, 2007.

Robbins, P. "Tracking Invasive Land Covers in India, or Why Our Land-
    scapes Have Never Been Modern." *Annals of the Association of Ameri-
    can Geographers* 91:637–659, 2001.

————. "Comparing Invasive Networks: Cultural and Political Biographies
    of Invasive Species." *Geographical Review* 94:139–156, 2004.

Roberts, S. "Barcoding Life." *Canadian Geographic,* March–April 2007.

Robertson, D. P., and R. B. Hull. "Public Ecology: An Environmental
    Science and Policy for Global Society." *Environmental Science & Policy*
    6:399–410, 2003.

Robinson, J. "Squaring the Circle? Some Thoughts on the Idea of Sustain-
    able Development." *Ecological Economics* 48:369–384, 2004.

Rogers, J. A. "Darwinism and Social Darwinism." *Journal of the History
    of Ideas* 33:265–280, 1972.

Rorty, R. *Contingency, Irony, and Solidarity.* Cambridge: Cambridge
    University Press, 1989.

Rose, M. R. *Darwin's Spectre: Evolutionary Biology in the Modern World.*
    Princeton, NJ: Princeton University Press, 1998.

Rosenberg, A. *Darwinism in Philosophy, Social Science and Policy.* Cam-
    bridge: Cambridge University Press, 2000.

Rosenthal, P. *Words and Values: Some Leading Words and Where They
    Lead Us.* New York: Oxford University Press, 1984.

Rosner, M., and T. R. Johnson. "Telling Stories: Metaphors of the Human
    Genome Project." *Hypatia* 10:104–129, 1995.

Ross, J. W. "The Militarization of Disease: Do We Really Want a War on
    AIDS?" *Soundings* 72:39–58, 1989.

Ross, N., J. Eyles, D. Cole, and A. Iannantuono. "The Ecosystem Health
    Metaphor in Science and Policy." *Canadian Geographer* 41:114–127, 1997.

Rowe, J. S. *Home Place: Essays in Ecology.* Edmonton, AB: NeWest Books,
    1990.

Rozzi, R. "The Reciprocal Links between Evolutionary-Ecological Sciences and Environmental Ethics." *BioScience* 49:911–921, 1999.

Rozzi, R., E. Hargrove, J. J. Armesto, S. Pickett, and J. Silander Jr. "'Natural Drift' as a Post-Modern Evolutionary Metaphor." *Revista Chilena de Historia Natural* 71:5–17, 1998.

Rubinoff, D. "Utility of Mitochondrial DNA Barcodes in Species Conservation." *Conservation Biology* 20:1026–1033, 2006.

Rue, L. *Everybody's Story: Waking Up to the Epic of Evolution.* Albany: State University of New York Press, 2000.

Rumelhart, D. E. "Some Problems with the Notion of Literal Meanings." In *Metaphor and Thought,* edited by A. Ortony, 2nd ed. Cambridge: Cambridge University Press, 1993.

Ruse, M. "Molecules to Men: Evolutionary Biology and Thoughts of Progress." In *Evolutionary Progress,* edited by M. H. Nitecki. Chicago: University of Chicago Press, 1988.

———. "Evolution and Progress." *Trends in Ecology & Evolution* 8:55–59, 1993.

———. *Monad to Man: The Concept of Progress in Evolutionary Biology.* Cambridge, MA: Harvard University Press, 1996.

Russell, E. P. "'Speaking of Annihilation': Mobilizing for War against Human and Insect Enemies, 1914–1945." *Journal of American History* 82:1505–1529, 1996.

Santa Ana, O. "'Like an Animal I Was Treated': Anti-Immigrant Metaphor in US Public Discourse." *Discourse and Society* 10:191–224, 1999.

———. *Brown Tide Rising: Metaphors of Latinos in Contemporary American Public Discourse.* Austin: University of Texas Press, 2002.

Sarewitz, D. R. *Frontiers of Illusion: Science, Technology, and the Politics of Progress.* Philadelphia: Temple University Press, 1996.

———. "How Science Makes Environmental Controversies Worse." *Environmental Science & Policy* 7:385–403, 2004.

Schiappa, E. "Towards a Pragmatic Approach to Definition: 'Wetlands' and the Politics of Meaning." In *Environmental Pragmatism,* edited by A. Light and E. Katz. New York: Routledge, 1996.

Schiebinger, L. *Nature's Body: Gender in the Making of Modern Science.* Boston: Beacon Press, 1993.

Schlaepfer, M. A., P. W. Sherman, B. Blossey, and M. C. Runge. "Introduced Species as Evolutionary Traps." *Ecology Letters* 8:241–246, 2005.

Schlanger, J. E. *Les métaphores de l'organisme.* Paris: J. Vrin, 1971.

Schön, D. A. *Invention and the Evolution of Ideas.* London: Tavistock, 1963.

———. "Generative Metaphor: A Perspective on Problem-Setting in Social

Policy." In *Metaphor and Thought,* edited by A. Ortony, 2nd ed. Cambridge: Cambridge University Press, 1993.

Schön, D. A., and M. Rein. *Frame Reflection: Toward the Resolution of Intractable Policy Controversies.* New York: Basic Books, 1994.

Schonlau, M., R. D. Fricker, and M. N. Elliot. *Conducting Research Surveys via E-Mail and the Web.* Santa Monica, CA: Rand, 2001, www.rand.org/pubs/monograph_reports/MR1480.

Schultz, B. "Language and the Natural Environment." In *The Ecolinguistics Reader: Language, Ecology and Environment,* edited by A. Fill and P. Mühlhäusler. New York: Continuum, 2001.

Schultz, P. W., and L. Zelezny. Reframing Environmental Messages to be Congruent with American Values. *Human Ecology Review* 10:126–136, 2003.

Schumacher, E. F. *A Guide for the Perplexed.* London: Abacus, 1977.

Schwartz, S. H. "Universals in the Context and Structure of Values: Theoretical Advances and Empirical Tests in 20 Countries." *Advances in Experimental Social Psychology* 25:1–65, 1992.

Sears, P. B. "Ecology—A Subversive Subject." *BioScience* 14:11–13, 1964.

Seed, J. "Beyond Anthropocentrism." In *Thinking Like a Mountain: Toward a Council of All Beings,* edited by J. Seed, J. Macy, P. Fleming, and A. Naess. Gabriola Island, BC: New Catalyst Books, 1988.

Shackleton, C. M., B. McGarry, S. Fourie, J. Gambiza, S. E. Shackleton, and C. Fabricius. "Assessing the Effects of Invasive Alien Species on Rural Livelihoods: Case Examples and a Framework from South Africa." *Human Ecology* 35:113–127, 2007.

Shanahan, T. "Evolutionary Progress?" *BioScience* 50:451–459, 2000.

Shapin, S., and S. Schaffer. *Leviathan and the Air-Pump: Hobbes, Boyle, and the Experimental Life.* Princeton, NJ: Princeton University Press, 1985.

Shepard, P., and D. McKinley, eds. *The Subversive Science: Essays toward an Ecology of Man.* Boston: Houghton Mifflin, 1969.

Shibles, W. *Metaphor: An Annotated Bibliography and History.* Whitewater, WI: Language Press, 1971.

Shrader-Frechette, K. S., and E. D. McCoy. "How the Tail Wags the Dog: How Value Judgments Determine Ecological Science." *Environmental Values* 3:107–120, 1994.

Sills, S. J., and C. Song. "Innovations in Survey Research: An Application of Web-Based Surveys." *Social Science Computer Review* 20:22–29, 2002.

Simberloff, D. "Confronting Invasive Species: A Form of Xenophobia?" *Biological Invasions* 5:179–192, 2003.

———. "Non-Native Species *Do* Threaten the Natural Environment!" *Journal of Agricultural & Environmental Ethics* 18:595–607, 2005.

———. "Invasional Meltdown 6 Years Later: Important Phenomenon,
    Unfortunate Metaphor, or Both?" *Ecology Letters* 9:912–919, 2006.
Simberloff, D., and B. Von Holle. "Positive Interactions of Nonindigenous
    Species: Invasional Meltdown?" *Biological Invasions* 1:21–32, 1999.
Singer, P. *A Darwinian Left: Politics, Evolution and Cooperation.* New
    Haven: Yale University Press, 1999.
Slobodkin, L. B. "The Good, the Bad and the Reified." *Evolutionary Ecology
    Research* 3:1–13, 2001.
Smart, N. *Worldviews: Crosscultural Explorations of Human Belief,* 3rd ed.
    Upper Saddle River, NJ: Prentice-Hall, 2000.
Smith, C. U. M. "Charles Darwin, the Origin of Consciousness, and
    Panpsychism." *Journal of the History of Biology* 11:245–267, 1978.
Smith, M. B. "The Metaphorical Basis of Selfhood." In *Culture and Self:
    Asian and Western Perspectives,* edited by A. J. Marsella, G. DeVos, and
    F. L. K. Hsu. New York: Tavistock Publications, 1985.
Sober, E. "Kindness and Cruelty in Evolution." In *Visions of Compassion:
    Western Scientists and Tibetan Buddhists Examine Human Nature,*
    edited by R. J. Davidson and A. Harrington. New York: Oxford
    University Press, 2002.
Solomon, S., D. Qin, M. Manning, Z. Chen, M. Marquis, K. B. Averyt,
    M. Tignor, and H. L. Miller, eds. *Climate Change 2007: The Physical
    Science Basis.* Contribution of Working Group I to the Fourth Assess-
    ment Report of the Intergovernmental Panel on Climate Change.
    Cambridge: Cambridge University Press, 2007.
Sontag, S. *Illness as Metaphor* [1978]; *and, AIDS and Its Metaphors.*
    New York: Doubleday, 1990.
Soulé, M. E., and G. Lease. *Reinventing Nature: Responses to Postmodern
    Deconstruction.* Washington, DC: Island Press, 1995.
Star, S. L., and J. Griesemer. "Institutional Ecology, 'Translations' and
    Boundary Objects: Amateurs and Professionals in Berkeley's Museum
    of Vertebrate Zoology, 1907–39." *Social Studies of Science* 19:387–420,
    1989.
Stepan, N. L. "Race and Gender: The Role of Analogy in Science." *Isis*
    77:261–277, 1986.
Stoeckle, M. Y., and P. D. N. Hebert. "Barcode of Life." *Scientific American,*
    October 2008, 82–88.
Stokes, K. E., K. P. O'Neill, W. I. Montgomery, J. T. A. Dick, C. A. Maggs,
    and R. A. McDonald. "The Importance of Stakeholder Engagement in
    Invasive Species Management: A Cross-Jurisdictional Perspective in
    Ireland." *Biodiversity and Conservation* 15:2829–2852, 2006.
Stoll-Kleemann, S., and T. O'Riordan. "From Participation to Partnership

in Biodiversity Protection: Experience from Germany and South Africa." *Society & Natural Resources* 15:157–173, 2002.

Swimme, B., and T. Berry. *The Universe Story: From the Primordial Flaring Forth to the Ecozoic Era—A Celebration of the Unfolding of the Cosmos.* New York: HarperCollins, 1992.

Szerszynski, B. "On Knowing What to Do: Environmentalism and the Modern Problematic." In *Risk, Environment and Modernity: Towards a New Ecology,* edited by S. Lash, B. Szerszynski and B. Wynne. Thousand Oaks, CA: Sage, 1996.

Takacs, D. *The Idea of Biodiversity: Philosophies of Paradise.* Baltimore: Johns Hopkins University Press, 1997.

Tanentzap, A. J., D. R. Bazely, P. A. Williams, and G. Hoogensen. "A Human Security Framework for the Management of Invasive Non-indigenous Plants." *Invasive Plant Science and Management* 2:99–109, 2009.

Taylor, P. J. "Technocratic Optimism, H. T. Odum, and the Partial Transformation of Ecological Metaphor after World War II." *Journal of the History of Biology* 21:213–244, 1988.

———. "Natural Selection: A Heavy Hand in Biological and Social Thought." *Science as Culture* 7:5–32, 1998.

———. *Unruly Complexity: Ecology, Interpretation, Engagement.* Chicago: University of Chicago Press, 2005.

Teilhard de Chardin, P. *The Phenomenon of Man.* New York: Harper and Brothers, 1959.

Thomas, L. *Late Night Thoughts on Listening to Mahler's Ninth Symphony.* 1983. Reprint, New York: Penguin, 1985.

Tobin, K., and D. J. Tippins. "Metaphors as Seeds for Conceptual Change and the Improvement of Science Teaching." *Science Education* 80:711–730, 1996.

Torgerson, D. *The Promise of Green Politics: Environmentalism and the Public Sphere.* Durham, NC: Duke University Press, 1999.

Toulmin, S. "The Construal of Reality: Criticism in Modern and Postmodern Science." *Critical Inquiry* 9:93–111, 1982.

———. *The Return to Cosmology: Postmodern Science and the Theology of Nature.* Berkeley and Los Angeles: University of California Press, 1983.

Trim, R. *Metaphor Networks: The Comparative Evolution of Figurative Language.* New York: Palgrave Macmillan, 2007.

Trombulak, S. C. "Misunderstanding Neo-Darwinism: A Reaction to Daly." *Conservation Biology* 14:1202–1203, 2000.

Trudgill, S. "Psychobiogeography: Meanings of Nature and Motivations for a Democratized Conservation Ethic." *Journal of Biogeography* 28:677–698, 2001.

Turbayne, C. M. *The Myth of Metaphor.* Columbia: University of South Carolina Press, 1970.

Turney, J. "The Sociable Gene—Finding a Working Metaphor to Describe the Function of Genes in an Organism Might Help to Ease Public Fears and Expectations of Genomic Research." *EMBO Reports* 6:808–810, 2005.

Ulanowicz, R. E. "Life after Newton: An Ecological Metaphysic." *Biosystems* 50:127–142, 1999.

Underhill, J. W. "The Switch: Metaphorical Representation of the War in Iraq from September 2002–May 2003," http://metaphorik.de/05/underhill.htm, 2003.

Underwood, R. A. "Toward a Poetics of Ecology: A Science in Search of Radical Metaphors." In *Ecology: Crisis and New Vision,* edited by R. E. Sherrell. Richmond, VA: John Knox Press, 1971.

Ungar, S. "Knowledge, Ignorance and the Popular Culture: Climate Change versus the Ozone Hole." *Public Understanding of Science* 9:297–312, 2000.

Uriarte, M., H. A. Ewing, V. T. Eviner, and K. C. Weathers. "Constructing a Broader and More Inclusive Value System in Science." *BioScience* 57:71–78, 2007.

Väliverronen, E. "Biodiversity and the Power of Metaphor in Environmental Discourse." *Science Studies* 11:19–34, 1998.

Väliverronen, E., and I. Hellsten. "From 'Burning Library' to 'Green Medicine': The Role of Metaphors in Communicating Biodiversity." *Science Communication* 24:229–245, 2002.

van Fraassen, B. C. "The World of Empiricism." In *Physics and Our View of the World,* edited by J. Hilgevoord. Cambridge: Cambridge University Press, 1994.

van Houtan, K. S. "Conservation as Virtue: A Scientific and Social Process for Conservation Ethics." *Conservation Biology* 20:1367–1372, 2006.

Venville, G. J., and D. F. Treagust. "Analogies in Biological Education: A Contentious Issue." *American Biology Teacher* 59:282–287, 1997.

Verhagen, F. C. "Worldviews and Metaphors in the Human-Nature Relationship: An Ecolinguistic Exploration through the Ages." *Language & Ecology* 2, 2008, www.ecoling.net/worldviews_and_metaphors_-_final.pdf.

von Bertalanffy, L. "An Essay on the Relativity of Categories." *Philosophy of Science* 22:243–263, 1955.

Wagar, W. W. *Good Tidings: The Belief in Progress from Darwin to Marcuse.*
     Bloomington: Indiana University Press, 1972.
Walker, B. "Ecosystems and Immune Systems: Useful Analogy or
     Stretching a Metaphor?" *Conservation Ecology* 5, 2001, www
     .ecologyandsociety.org/vol5/iss1/art16.
Walker, B., and D. Salt. *Resilience Thinking: Sustaining Ecosystems and
     People in a Changing World.* Washington, DC: Island Press, 2006.
Wallace, B. A. *The Taboo of Subjectivity: Toward a New Science of
     Consciousness.* New York: Oxford University Press, 2000.
Wallington, T. J., and S. A. Moore. "Ecology, Values, and Objectivity:
     Advancing the Debate." *BioScience* 55:873–878, 2005.
Ward, B. *Spaceship Earth.* New York: Columbia University Press, 1966.
Webb, T. J., and D. Raffaelli. "Conversations in Conservation: Revealing
     and Dealing with Language Differences in Environmental Conflicts."
     Journal of Applied Ecology 45:1198–1204, 2008.
Weber, J. R., and C. S. Word. "The Communication Process as Evaluative
     Context: What Do Nonscientists Hear When Scientists Speak?"
     BioScience 51:487–495, 2001.
Weingart, P. "'Struggle for Existence': Selection and Retention of a Meta-
     phor." In *Biology as Society, Society as Biology: Metaphors,* edited by
     S. Maasen, E. Mendelsohn, and P. Weingart. Dordrecht: Kluwer, 1995.
Weingart, P., A. Engels, and P. Pansegrau. "Risks of Communication:
     Discourse on Climate Change in Science, Politics, and the Mass Media."
     Public Understanding of Science 9:261–283, 2000.
Weingart, P., and S. Maasen. "The Order of Meaning: The Career of Chaos
     as a Metaphor." *Configurations* 5:463–520, 1997.
Westoby, M. "What Does 'Ecology' Mean?" *Trends in Ecology & Evolution*
     12:166, 1997.
Wheeler, Q. D. "Losing the Plot: DNA 'Barcodes' and Taxonomy." *Cladistics*
     21:405–407, 2005.
Whitehead, A. N. *Science and the Modern World.* New York: Macmillan,
     1925.
Whitmarsh, L. "What's in a Name? Commonalities and Differences in
     Public Understanding of 'Climate Change' and 'Global Warming.'"
     Public Understanding of Science 18:401–420, 2009.
Whorf, B. J. *Language, Thought, and Reality.* Cambridge, MA: MIT Press,
     1956.
Wicklum, D., and R. W. Davies. "Ecosystem Health and Integrity?"
     Canadian Journal of Botany 73:997–1000, 1995.
Wilkinson, D. "The Parable of Green Mountain: Ascension Island,

Ecosystem Construction and Ecological Fitting." *Journal of Bio-geography* 31:1–4, 2004.

Williams, B. *Ethics and the Limits of Philosophy*. Cambridge, MA: Harvard University Press, 1986.

Williams, G. C. *Adaptation and Natural Selection*. Princeton, NJ: Princeton University Press, 1966.

Williams, R. *Keywords: A Vocabulary of Culture and Society*. New York: Oxford University Press, 1983.

Wilsdon, J., and R. Willis. *See through Science: Why Public Engagement Needs to Move Upstream*. London: Demos, 2004.

Wilsdon, J., B. Wynne, and J. Stilgoe. *The Public Value of Science: Or How to Ensure That Science Really Matters*. London: Demos, 2005.

Wilson, D. S., E. Deitrich, and A. B. Clark. "On the Inappropriate Use of the Naturalistic Fallacy in Evolutionary Psychology." *Biology and Philosophy* 18:669–682, 2003.

Wilson, E. O. *Naturalist*. Washington, DC: Island Press, 1994.

Winawer, J., N. Witthoft, M. Frank, L. Wu, A. Wade, and L. Boroditsky. "Russian Blues Reveal Effects of Language on Color Discrimination." *Proceedings of the National Academy of Science* 104:7780–7785, 2007.

Winner, L. *The Whale and the Reactor: A Search for Limits in an Age of High Technology*. Chicago: University of Chicago Press, 1986.

Wittgenstein, L. *Philosophical Investigations* (1953), translated by G. E. M. Anscombe, 3rd ed. Malden, MA: Blackwell, 2001.

Woodhouse, E., D. Hess, S. Breyman, and B. Martin. "Science Studies and Activism: Possibilities and Problems for Reconstructivist Agendas." *Social Studies of Science* 32:297–319, 2002.

Worster, D. *Nature's Economy: A History of Ecological Ideas*. Cambridge: Cambridge University Press, 1985.

———. "The Ecology of Order and Chaos." *Environmental History Review* 14:1–18, 1990.

Wright, S. "Panpsychism and Science." In *Mind in Nature: Essays on the Interface of Science and Philosophy*, edited by J. B. Cobb Jr. and D. R. Griffin. Washington, DC: University Press of America, 1977.

Wu, J., and O. L. Loucks. "From Balance of Nature to Hierarchical Patch Dynamics: A Paradigm Shift in Ecology." *Quarterly Review of Biology* 70:439–466, 1995.

Wynne, B. "May the Sheep Safely Graze? A Reflexive View of the Expert-Lay Knowledge Divide." In *Risk, Environment and Modernity: Towards a New Ecology*, edited by S. Lash, B. Szerszynski, and B. Wynne. Thousand Oaks, CA: Sage, 1996.

Yearley, S. *Sociology, Environmentalism, Globalization: Reinventing the Globe.* Thousand Oaks, CA: Sage, 1996.

Yoon, C. K. *Naming Nature: The Clash between Instinct and Science.* New York: W. W. Norton, 2009.

Young, N., and R. Matthews. "Experts' Understanding of the Public: Knowledge Control in a Risk Controversy." *Public Understanding of Science* 16:123–144, 2007.

Young, R. M. "Darwin's Metaphor: Does Nature Select?" *Monist* 55: 442–503, 1971.

———. "Darwinism *Is* Social." In *The Darwinian Heritage,* edited by D. Kohn. Princeton, NJ: Princeton University Press, 1985.

———. *Darwin's Metaphor: Nature's Place in Victorian Culture.* Cambridge: Cambridge University Press, 1985.

———. "Darwin's Metaphor and the Philosophy of Science." *Science as Culture* 3:375–403, 1993.

Zaehner, R. C. *Evolution in Religion: A Study in Sri Aurobindo and Pierre Teilhard de Chardin.* Oxford: Oxford University Press, 1971.

Zimmerman, C., and K. Cuddington. "Ambiguous, Circular and Polysemous: Students' Definitions of the 'Balance of Nature' Metaphor." *Public Understanding of Science* 16:393–406, 2007.

Zuk, M. "Feminism and the Study of Animal Behavior." *BioScience* 43:774–778, 1993.

Zwicky, J. *Wisdom and Metaphor.* Kentville, NS: Gaspereau Press, 2003.

Index

progress: contrasting views of, 34–
40; historical background, 32–
34; as measure of sustainability,
48–55; progress metaphors, 25–
26, 32–33, 40–42; and sense of
purpose, 56–57, 65; types of, 35–
40, 45f
promotional metaphors, 133,
134–135
public expertise, 204
purpose, sense of: evolutionaries'
views on, 60; evolutionary biologists' views on, 59–60; and
progress, 56–57, 65
Putnam, Hilary, 90

Ravetz, Jerome, 12, 157, 209–210
Rawles, Kate, 161
referential adequacy, 95
relative progress metaphors, 50–51
religion and science, 97–99,
116–117
resonance of metaphors, 12–14
Richards, Robert, 37
Richardson, David, 186, 187
Rilke, Rainer Maria, 223
Ross, J. W., 161
Rozzi, Ricardo, 87
Ruse, Michael, 33, 35–36

Sahtouris, Elisabet, 63
Sandstrom, Kent, 129–130
Santa Ana, Otto, 109
Sarewitz, Daniel, 155, 199
Schön, Donald, 11
Schwartz, Shalom, 78
science: deficit model of public
knowledge, 13–14, 219; ethics of,
94, 218–219; facts and values relationships, 69–75; and metaphors, 3–4, 6–7, 22–23, 63–65;

personification of nature, 58–59;
politics in, 131–132, 199; progress
debate, 53–54; and religion, 97–
99, 116–117; as value-free, 71–72,
88; values and, 73–75, 89–90, 199.
See also communication, science;
scientists; society and science
scientification of politics, 131–132
scientific progress metaphors, 53
scientists: and advocacy, 21–22,
149, 221–222, 225–226; on competition in nature, 83–87; credibility of, 171–172; funding, 132–
133; language use, 95–98, 219;
media relations, 174; metaphor
development, 18, 74f, 90–91,
128–133, 216–219; policy development, 21–22; power position
of, 130–133, 135, 145–146, 167, 173,
174; promoting values, 28–29;
roles of, 11–12, 20
selfish gene metaphor, 131–132
Simberloff, Dan, 165–168, 169,
174–175
Sober, Elliott, 82, 85
socially adequate language, 96
social marketing with metaphors,
205
society and science: contextual
interpretations, 15–16; facts and
values struggle, 90–91; metaphorical links, 6–9, 227–228;
need for interaction between,
11–12; politics in, 198; progress
metaphors, 25–26, 32–34, 34–40;
public participation, 100; recognizing expertise, 198–199; relation to religion, 97–99, 116–117.
See also values
Society for the Study of Evolution
(SSE), 34–40